HOW TO RAISE CATTLE

EVERYTHING YOU NEED TO KNOW

BREED GUIDE & SELECTION
PROPER CARE & HEALTHY FEEDING
BUILDING FACILITIES AND FENCING
SHOWING ADVICE

Philip Hasheider

Voyageur Press

DEDICATION

To my wife, Mary, whose constant support and encouragement helped bring this book to a successful conclusion.

First published in 2007 by MBI Publishing Company and Voyageur Press, an imprint of MBI Publishing Company, 400 1st Avenue North, Suite 300, Minneapolis, MN 55401 USA

The contents of this book were reviewed and approved by Dr. Clint Rusk, Associate Professor of the Youth Development and Ag. Education Department at Purdue University, in accordance with industry standards.

Library of Congress Cataloging-in-Publication Data

Hasheider, Philip H., 1951-
How to raise cattle : everything you need to know / Philip H. Hasheider.
p. cm.
Includes index.
ISBN: 978-0-7603-2802-6 (softbound)
1. Cattle. 2. Beef cattle. 3. Dairy cattle. 4. Ranching. 5. Dairy farming. I. Title.
SF197.H37 2007
636.2—dc22

2006024854

Editor: Amy Glaser
Designer: LeAnn Kuhlmann

To find out more about our books, join us online at www.voyageurpress.com.

Printed in China

About the Author
Philip Hasheider lives on a farm with his wife and children in rural Sauk City, Wisconsin.

CONTENTS

ACKNOWLEDGMENTS

Two men—Howard Hasheider, my father, and William A. Weeks, an astute stockman and colleague—deserve the most credit for teaching me how to look at cattle differently, in ways not taught by textbooks. Without their unique insights, advice, guidance, and patience, my ability to understand cattle and how to breed and raise them would be greatly diminished.

I particularly wish to thank Jerry and Ruth Apps, Dr. Helen Hotz, Petrina Green, and Beverly Davidson for supporting my writing efforts and offering continuing encouragement.

My son, Marcus, spent many hours with me taking photographs and was an enthusiastic supporter through this whole project. My wife, Mary, read several drafts of this book, each time offering comments and suggestions that clarified points I wished to make. My daughter, Julia, was a constant support and her indomitable teenage spirit always brightened a difficult moment.

Others whose help with this book was greatly appreciated include Dr. Robert Bremel; Ray and Linda Brickl; Mike Gingrich; Henk, Johanna, Hans, and Dorthe Griesen; John Hasheider; Dr. Jane Homan; John Kiefer; Katherine and Jeff Lohr; Tom Miller; Robert and Donna Ochsner; Randy Sprecher; William Stade; Eldon Ward; Robert Wills; and Fritz and Ginny Wyttenbach.

Also, a thank you to my editor, Amy Glaser, whose farm background allowed her to see the positive impact a cattle-raising book could have for readers.

531 56 152
198

CHAPTER 1

AN INTRODUCTION TO RAISING CATTLE

Contented animals are the result of a well-thought-out farming program where each animal has the opportunity to grow in a humane, environmentally friendly manner.

A farm is a place where you can feel the sun on your face, the wind blowing in your hair, and the freshness of each day. No two days on a farm are typically the same. Each day comes fresh with a new set of circumstances to be addressed and challenges to be solved.

By choosing to live on a farm and raise cattle, you'll learn many things about yourself. You'll learn about your resourcefulness, your attitudes about life, your family, and what you hold most dear. Farming is about creation: taking seeds and growing crops, raising cows to have calves, stewardship of the land, and re-creating something anew. Whether raising a herd of beef or dairy animals, you're setting forth the process of creation and doing it with living

With the end of the day comes a time for reflection. Your decision to farm and raise cattle can give you much joy, pride, and a deep sense of fulfillment.

Family farms have been the cornerstone of agriculture's success in the United States for generations. By becoming a part of this tradition, you will also play a part in its future success.

Raising beef cattle brings the full circle of life to your farm. Each year a new crop of calves arrives to restock your pastures. As the females grow and mature, they produce the next generation, helping you to continue building your business.

beings. It is a power that few people have the privilege to learn and understand, let alone experience. If raising cattle is your goal, this book will help you discover the things that will prepare you for success and enjoyment as you work with your animals.

PLANNING YOUR ENTERPRISE

Before jumping headlong into your dreams of wide-open pastures and viewing your spread from a split-rail fence, it is worth your time to think about and consider the reasons you'd like to raise cattle.

A simple exercise can clarify your objectives and give you a greater chance for success with your enterprise. List all the reasons you think you have for wanting to raise cattle. When you have finished, put them in order of importance, beginning with your three main reasons.

This is not a trivial exercise. You need to understand your reasons for raising cattle if you are to succeed. Developing and stocking a beef or dairy farm can be a significant financial investment and this alone is reason enough to ask yourself some questions such as: Are you willing to commit the time necessary to complete the day-to-day work load? And, do you have a desire to handle animals and general farm work?

Understanding traditional country values will enhance your rural experience. Being friendly, a good neighbor, and respectful of others and their property are all pieces of becoming part of the social fabric within the community.

ANIMAL HUSBANDRY AND CONSUMERS

Developing a beef or dairy herd with the goal of marketing animals for meat or producing milk and dairy products can be financially rewarding. Home-grown meat from farms that are often referred to as "family farms" is a growing niche market. Producing quality meat or milk and dairy products from animals that are raised in a humane, environmentally friendly manner can give consumers—whether they purchase directly from you or through a system that markets products from other farmers like you—a sense of connection and well-being.

Leaving an urban environment to live in a rural setting can be an exciting and rewarding decision for you and your family. Moving to a rural area also brings certain responsibilities, such as land stewardship and helping maintain the rural character for future generations.

More and more of today's consumers are seeking high-quality food from animals grown under wholesome conditions that can in some way connect these consumers to the land. Perhaps these consumers sense something in their background, heritage, or belief system that draws them to that reason, although in reality they may be far removed from agriculture. Consumers have their own reasons for seeking this connection just as you have your reasons for wanting to raise cattle. This connection can be a vital link in your business success and will provide you an opportunity to develop and use your animal husbandry talents.

Some consumers may feel a sense of attachment to the rural character when buying from those who produce food in small-scale enterprises. Call it a back-to-the-earth mystique, but it can be a strong pull that can provide you with a ready market. If one of the reasons you listed for wanting to raise cattle is to make money, then tapping into this market can have substantial benefits for you and your family. Value-added products are an avenue to make the most of your investment and are discussed in chapter 14.

Town government is the basis for our society, and monthly town meetings provide residents an opportunity to introduce and discuss local issues of concern. By participating in this process, you can immerse yourself in the community's affairs and meet others who live in your township.

Involving the entire family in farming can help family members develop unique skills, positive attitudes, and responsibilities that can last a lifetime. Interest in the cattle is heightened when everyone has a stake in the outcome and welfare of the animals being raised.

FARMING PRACTICES AND CATTLE

Perhaps you desire to work with animals and grow them under the most natural environment possible, or to create a farm that uses traditional farming practices. Animals grown under grass-based farming methods used 40 and 50 years ago yield meat products that studies have shown may be more nutritious for consumers than cattle grown under today's intensive, large-scale farming systems.

Animals allowed to mature at a rate traditionally considered sufficient for making a profit have a different feeding protocol than those fed for today's accelerated growth rates. This difference may be an important consideration in your farming decision.

A slower, less-aggressive growth rate is more in tune with the natural rhythm of the animal's body. It keeps the cattle closer to the pattern that nature long ago established for the development and maturation of the animal. Producers who believe in this natural growth pattern reject the systematic use of growth hormones, implants, and supplemental feed ingredients, all of which stimulate an animal's appetite and encourage it to ingest larger quantities of feed over a shorter time. One of your reasons for raising beef cattle may be the pleasure you experience from assisting an animal's development in the most traditional way possible.

Tapping into the consumers' sense of how this rhythm of natural growth can benefit their families can also be one satisfying reason for deciding to raise cattle. You then also become a part of this natural rhythm of farming.

This wholesomeness of small-scale animal production may be another reason that raising cattle can be attractive to you and your customers. The image of wide-open spaces, sunlight, freedom of movement, and green pastures can be a strong tug at customers' emotions. All these terms envision an animal that lives, sleeps, breathes, eats, and grows up under the sun and the stars. This powerful image can be

As more consumers become aware of food safety issues, the number of people seeking locally grown, quality food increases. Becoming a part of a system that delivers food products desired by consumers can be personally and financially rewarding.

cultivated to your advantage and can also provide you with a deep sense of accomplishment at the end of the day.

DAIRYING IS DIFFERENT THAN BEEF PRODUCTION

If you are thinking about starting a dairy farm in order to milk cows, you will have different considerations than a person who wants to raise cows for beef, such as herd size and the market you wish to pursue. Perhaps you want to produce only enough milk for your family or to supply a small local market. Maybe you're thinking on a bigger scale and want to have a large dairy farm so you can produce great quantities of milk.

Understanding your reasons for dairying will determine the financial investment necessary and increase your chances for success. A dairy farm has a greater financial investment than a farm used for raising beef cattle because of the high cost of dairy cows, buildings, the milking system, and machinery. Dairying also has an extensive time commitment since cows are milked twice a day, every day.

The current dairy economy can accommodate both small and large herds. Large dairies use economies of scale, meaning the investment is spread over a larger number of animals to help manage costs. Small dairies can compete and be viable if they are able to keep their costs of production at the lowest possible level. Grass-based, sustainable, and organic dairy farms are expanding in number because of this potential, and this allows entry into the industry by those who do not have significant financial assets. Milk produced through these systems attracts the artisan cheese venues that produce a variety of unique cheeses for varying consumer tastes and has experienced an expanding market share.

Those who pursue these niche dairy markets can offer a food product similar to those raising beef animals: a wholesome product that is produced in the most natural way possible.

LIFESTYLE AND LIVING ON THE LAND

If you are considering a move from an urban setting to a rural area, already have a connection to a farm from your past and want to go back to your roots, or desire a lifestyle change, then raising cattle may help you accomplish and enhance these goals.

In years past, it was unusual for a nonfarmer to leave a town or city to take up the pursuit of farming. Lack of experience or money often served as roadblocks to that pursuit. However, this is changing as the pressures of urban living, along with what some sense to be a lack of fulfillment in their lives, steer more people to contemplate other options for making a life and a living.

If you live in a city or town, it's likely that you commute to your place of employment. On a farm, your job is before you the minute you step out the door. You're surrounded by your new life. There is no commute or traffic jam as you walk across the farmyard to go work with your animals.

Owning and running a small-scale farm is like operating any small business where few, if any, employees outside the family are needed. Operating a large-scale farm is very much like running a large-scale business in a town or city because both are capital- and management-intensive. Managers of large farms need expertise in hiring and managing qualified employees and the ability to train them to successfully complete the task at hand.

In most areas of the United States, grass grows where mechanical harvesting is absent or impossible. Hills and low areas not useable for any other crops can be efficiently used by cattle in pasture-grazing programs.

There are opportunities and challenges that exist whether you're thinking small or large. If you have the skills for handling several employees, then you may wish to consider developing a large-scale operation, providing you have sufficient capital to invest.

If you simply want to own land and have animals walking on it, the purchase of a farm as an investment can be a consideration. Land is a finite resource and for that reason it's always in demand. Land is a physical resource and brings with it the responsibility and challenge of preserving and utilizing it as best you can. Land ownership entails stewardship and keeping it viable for future generations.

Perhaps you already own a farm and are thinking about purchasing some animals to roam your pastures or fields that are difficult to cultivate. As ruminant animals, cattle are natural grazers and have the ability to forage in areas where mechanical harvesting equipment cannot be used.

FAMILIES AND FARMING TOGETHER

If you'd like to raise cattle and involve other family members in your enterprise, you'll find there are many valuable lessons to be learned by everyone. Family involvement can lead to learning and accepting responsibility, as well as developing

Feelings of self-satisfaction and a sense of accomplishment can be part of your reflections at the end of each day. What you've learned and experienced can never be taken from you.

an emotional satisfaction that comes from caring for animals that depend on you for their own well-being. Farming develops self-sufficiency as historically farmers have relied on themselves to accomplish many of their goals. Self-sufficiency can be a fulfilling life experience and expanding one's limits can be rewarding for you and your family.

The very nature of farming involves many members of the family because of the workload involved in caring for animals. This involvement has expressed itself in many ways down through the years but it has led to a general sense of stability as farms were handed from generation to generation.

As a farmer, you will enter that circle and may feel that same sense of stewardship that millions of farmers before you have experienced. The percentage of people alive today who can identify with that experience is very small and you will share a part of that distinctive community.

Just as the animals have a rhythm in which they grow, land and nature also have a rhythm of their own. It is a rhythm that does not require the presence of humans but where humans must find their presence in order to benefit from the gifts of the land.

Small rural communities are nestled within farming areas and serve as anchors for farm families, as well as centers for many social activities.

CHAPTER 2

HUMANS AND CATTLE— A HISTORICAL PERSPECTIVE

The relationship between humans and cattle extends back to some of the earliest recorded times. Remains of domesticated cattle have been found in Turkey and other sites in Eastern Asia that date to 6500 B.C. The eventual domestication of cattle provided meat and milk to supplement the diet of their keepers and a power source for labor. Hunting and taming these animals provided everything from hides, which were fashioned into clothing and foot wear, to dung, which was used for cooking and building.

For centuries, cattle provided a measure of wealth, and even today, for some cultures such as the Maasai of Africa, cattle ownership remains a sign of wealth and is central to an economy where animals are traded or sold to settle debts.

Early humans are reported to have both feared and respected cattle because of their crescent-shaped horns, which appeared to have had religious significance. Today, in certain parts of the world, cattle still have a major religious role, particularly in Hinduism where the cow is

You can feel on top of the world by moving from an urban area to a rural setting. For generations, farmers have had a unique bond with land and animals.

All breeds of domestic beef and dairy livestock are descended from the aurochs. Nature's many bovine varieties and evolutions have provided humans with many breeds to choose from.

believed to be holy and a divine cow named Kamadhenu is considered the mother of all Hindu gods.

ANCESTORS OF MODERN-DAY CATTLE

Carolus Linneaus, often called the father of taxonomy, classified cattle into three separate species. *Bos taurus* identified the European cattle, including similar types from Africa and Asia. *Bos indicus* included the zebu. The now-extinct *Bos primigenius* identified the aurochs.

Aurochs, the wild ancestor of European cattle, were hunted for their meat. Prehistoric paintings substantiate the appearance of the auroch to early man. The Auroch was very large in size and survived until relatively modern times. The last surviving member is reported to have died in Poland in 1627. Today's domestic cattle evolved from the Aurochs.

Zebu cattle are related to the Aurochs and were domesticated in India over 10,000 years ago. They have large humps on their backs. The majority of zebu breeds today are located in Africa and South Africa.

Contemporary purebred and commercial breeding programs have benefited from generations of dedicated cattle producers whose lives were spent improving the quality of their stock.

Cattle domestication greatly enhanced early man's diet. Milk provided a nutritious drink and meat provided protein. Having cattle close at hand helped alleviate the need for a daily hunt for food. The ease in which animals could be fed added to their popularity. Cattle are ruminants with a unique digestive system that allows them to digest plants otherwise unpalatable to humans. Grazing cattle could readily turn grass and plants into a source of food that humans could better utilize.

As human development progressed, cattle became confined to small areas of individual farms. Developing different breeds of beef and dairy cattle as we know them also started around this time.

Today there are about 260 different cattle breeds, types, and varieties in different parts of the world. Interestingly, this includes the buffalo and bison of North America and the yak in China.

CATTLE BREED DEVELOPMENT

Robert Bakewell, an English livestock breeder who lived in the 1700s, is credited with being the first to develop specific cattle breeds. He used the idea of mating animals with similar characteristics until those patterns became fixed or established within the offspring. Bakewell's great leap of understanding was to keep cattle from breeding at random, as had been done previously. Instead, he separated the males and females and allowed mating to occur only deliberately and specifically. This allowed him to fix and enhance those traits he felt to be desirable.

Bakewell developed the Dishley, or New Leicestershire, longhorn cattle breed. Although few cattle today are based on the breeds with which Bakewell worked, his cattle breeding methods have become an accepted practice worldwide.

Raising cattle can be aesthetically pleasing, as well as emotionally and financially rewarding. Working with cattle can take you into another world where you learn about animal behavior, personalities, and purposeful lives.

Grassland birds benefit from pasture-grazing programs. Mechanical harvesting discourages birds from nesting in fields where hay crops are taken off early in the season, but pastures where cows are allowed to graze allow birds time to nest and raise their young. The insects scattered as cows walk about are a food source for the birds.

Pasture-grazing hillsides make good use of marginal land or land that is not suitable for tilling because of potential erosion. Whether using grass in a beef- or dairy-grazing program, land benefits from animals being on it.

BEEF BREEDS IN THE UNITED STATES

European emigrants brought beef cattle to the United States during the mid- to late 1700s and this continued well into the 1800s. The first importations were generally from Great Britain where numerous breeds were being developed. The Angus (Black and Red), Hereford (Horned and Polled), and Shorthorn were brought from England at that time and are among the primary beef breeds raised in the United States today.

During the late 1960s and early 1970s major imports of continental European breeds were introduced and made available to American beef producers. Commonly referred to as exotic breeds, these included Charolais, Chianina, Gelbvieh, Limousin, Maine Anjou, Simmental, and others. There are over 60 beef breeds represented in the United States today. However, about 20 breeds constitute the majority of the genetics that are used for commercial beef production. See Chapter 6 for more complete information.

DAIRY BREEDS IN THE UNITED STATES

European dairy cattle arrived in the United States as early as 1611 at the Jamestown Colony in Roanoke, Virginia. It was another 200 years before an effort was organized to import cattle used for the specific purpose of producing milk for consumption.

By the late 1800s major imports of dairy cattle to our country began in earnest. Early importers brought Holstein-Friesians from Holland and Northern Germany, Brown Swiss from Switzerland, Jerseys and Guernseys from the British Channel Islands, and Ayrshires from Scotland. Other minor breeds entered the country over the next 25 years. Importations of European dairy cattle to the United States effectively ceased in 1906 after an outbreak of foot and mouth disease spread across Europe. Today there are six major dairy breeds and a half dozen less-prevalent breeds used for milk production in the United States. See Chapter 17 for more information.

A beef herd does not need to be large in number to provide meat for you and your family. Cattle can utilize areas difficult to mechanically harvest.

CHAPTER 3

GETTING STARTED—FARM AND CATTLE

Starting a cattle-raising business requires land, buildings, animals, and equipment. When you purchase a farm, you are usually purchasing a business as well. Besides managing the physical assets of the farm, you will also need to acquaint yourself with the business practices associated with farming.

Good planning and research, and obtaining good advice will help you avoid unpleasant surprises. Advice can come from an agriculture lending group or bank, a county agricultural extension office, or private professional services that specialize in farm purchases and setting up farming enterprises. You can do much of the initial research on your own by contacting real estate agents about the availability of farms for sale or rent, or by visiting properties on your own.

Farming offers a unique way of life, as well as a rewarding business. Starting with a farm that is right for your situation will give you a greater chance for success and a deep sense of pleasure and accomplishment.

FARMS AND PROPERTY ACQUISITION

Buying a farm is a large financial investment so choose your farm carefully and get professional advice if you are unsure how to proceed. When purchasing a farm, you may have a strong attraction to live in a certain area or region because of climate considerations, soil type, or the distance from a major city or village. In some cases, the appeal may be the distance away from population centers.

Renting a farm for a period of time may help you ease into cattle raising as well as help you position your finances toward an eventual purchase. By renting, you may be able to determine the farm's productive capacity and the workability of the facilities available (such as the house, barn, sheds, and fences).

LOCATION AND SOCIAL CONSIDERATIONS

In many cases your farm will also be your home. The property's location and the services available may be important factors when deciding which property to buy. Living in a rural area is not the end of the world; however, there are some significant geographic differences between rural settings and urban ones. Living on a farm does not necessarily exclude you or your family from the conveniences or services available in urban areas. There just happens to be a greater distance to access them.

When purchasing a farm, assess whether the house or dwelling meets your family requirements both now and in the future. If you have a young family, it may be important to be located close to schools, doctors, or transportation systems. The availability of professional veterinary services is important for your animals. If community activities are important to you and your family, you can visit the area's chamber of commerce, which provides information about local activities during the year. You may also want to look at the opportunity for alternative or off-farm income or employment.

Purchasing a farm opens many new possibilities for you and your family. Living close to the land can be a financially rewarding and emotionally satisfying experience.

PHYSICAL CONSIDERATIONS AND APPLICATION

When buying a farm, the physical factors you are considering, such as location and climate, may be closely tied to your social considerations. Farm size is an important aspect of a property and will determine the number of cattle you can raise, whether they are beef or dairy cattle, your ability to spread the manure produced by your animals, and other factors.

Soil type and fertility are important considerations because in some cases the soil type is tied to the value of the property. Soil type can also influence the types of crops raised and the durability of the crops during a drought or extended dry spell. Heavy soils help sustain crops in dry conditions while lighter, sandier soils do not. Another aspect to consider is the past fertilizer history. This is especially important if you are interested in starting an organic certified farm.

The quality of the buildings and of improvements made to them can be a determining factor in a farm purchase. Extensive building renovations may require finances that could otherwise be directed toward the operating expenses of your farm. Yet, the need for these same improvements may lower the purchase price and be an attractive option.

Whatever farm you purchase, it is necessary to fully understand the boundaries. Walking the fences will provide you with a better idea of how much land is there as well as information about the condition of the fences, the buildings, and the soil and other aspects of the property such as if a grazing program is possible or not.

Before purchasing a property, be sure to check for the possible presence of contaminants or residues that could affect the health of your family or animals. These may be difficult to identify because conventional farming has typically used chemicals on crops. Underground fuel storage on farms has been banned in many states. However, old storage tanks may still be present and need to be dug up and removed. Assess the history of the property you are considering. Be sure to address this issue prior to signing any purchase agreement.

Good buildings are a sound investment and will withstand many years of use. Sound structures are safe and need minimal maintenance, which will allow more monies to be available for operating expenses.

FINANCING AND CATTLE PURCHASES

Purchasing a farm requires many financial considerations, including creating a budget for your lending source and providing capital costs for initial purchases such as fertilizers, animals, and equipment. Legal advice and assistance should be obtained before entering into any purchase commitments.

If you do not already own cattle, you will need to find a reliable source where you can purchase the animals. There are a number of factors to consider when purchasing cattle. Some may be self-evident, while others may take more research on your part. If you are uncertain about how to purchase a beef or dairy animal, it's best to work with someone who is ethical and understands this side of the business.

Some general considerations when purchasing animals include their overall physical condition, health, mobility, frame size, and breed. While these are not necessarily the only considerations, they will provide a foundation for cattle selection.

Pasture corrals are easy to build. They help contain cattle when loading and unloading, and when capturing cattle that need attention. Allowing daily access to this area will familiarize animals with the corral so they will not be afraid when it is used.

Purchasing healthy animals when you start farming is the best way to maintain a healthy herd over time. Healthy animals are easier to raise because they grow better by utilizing feed more efficiently, and they cost less to maintain.

1. **Condition** refers to the amount of weight or fat the animal carries on its body. Animals that will be pastured should not be excessively fat because they will not gain weight very well during the first month as they make the transition to your farm. Cattle should be healthy and lean. If possible, avoid animals that have not performed well elsewhere. They may be less expensive but they won't achieve an acceptable level of performance.
2. **Health** is the animal's general physical condition. Animals you purchase should have a bright hair coat and not a dull, emaciated look. Normal breathing should be observed. Avoid animals with runny noses or the lack of a bright eye. Try to locate any ear vaccination tags as this will help indicate the care they have received from the time they were young calves. Farmers who vaccinate their heifer calves for brucellosis or other viruses are usually concerned with the health of the animals.
3. **Mobility** is another health consideration. Simply watch the animals move around. An animal should have the ability to move about freely with no leg, joint, or feet problems. Avoid animals that limp and have swollen joints, long hooves, or other physical impairments. Animals exhibiting any of these conditions will not last long on any farm.

4. **Frame size** is especially important if you are raising your animals for market. Cattle with a small frame size will finish at a light weight, while cattle with a large frame will generally finish at a heavier weight because they can carry more flesh over a larger skeleton. If you plan to manage your cattle as a single group, purchase animals uniform in frame size.
5. **Breed** is usually not as important as the frame and conformation of the cattle. Many of the beef breeds are populated with animals of similar frame sizes. Animal disposition is also important and there are differences between breeds. Try to avoid aggressive or high-strung cattle.

Purchasing pregnant cows will give you a quicker start to your program than purchasing non-pregnant cows. If waiting is not a problem, it may be less expensive to buy non-pregnant cows and breed them yourself. However, be aware that buying an open, or nonpregnant, cow may be more costly in the long run. Not being pregnant may be an indication that there is a problem with the reproductive health of the animal. Cows that fail to become pregnant are investments that result in lost income potential.

If you purchase one or several cows at a time it is best to have them pregnancy checked by a veterinarian before buying them. Consider making the results of this pregnancy check part of the purchase agreement.

Custom-hire operators can harvest your crops so you don't have to invest in large amounts of equipment. Timely harvesting helps attain the plant's highest nutrient content, which will increase the quality of the feed for your cattle's optimum growth.

BUILDINGS AND EQUIPMENT

The facilities needed for raising cattle do not need to be extensive but they may determine to some extent the size of your program and the number of animals with which you can easily work. Buildings do not need to be elaborate or expensive but they do need to be sturdy enough to withstand the movement of cattle and the pressures their body types exert upon them. Most of the discussion of buildings and other facilities will be taken up in chapter 4.

Raising crops to feed your cattle will require equipment to help with the harvest of hay, corn, or any other forage you grow. This applies whether you own or rent your farm. In some regions the climate is moderate enough that a winter feed supply is available and can be supplemented with stored feed. However, in most northern areas of the country, a winter feed supply must usually come from stored feed harvested during the growing season. To maintain a sufficient feed supply for your cattle, you will have to harvest crops during the growing season or purchase all of your cattle's nutrition needs.

You must first decide whether you will do all or most of the harvesting yourself or if you want to hire the work done for you. Then determine what kind of equipment you need to efficiently complete a harvest of hay, grass, corn, or whatever else you decide to use as the major source of roughage for your cattle.

If you decide to do the field work yourself, you will need to purchase quite a bit of equipment. Tractors, wagons, hay cutters, balers, forage choppers, and tilling machinery are only some of the equipment required. It is very important that you, as well as family members and employees, have sufficient experience in safely and efficiently operating

Most farms have many tools and implements that are used daily. By attending a local farm auction, you will be able to better understand some of the items you may find useful.

A tractor, a chopper, and wagon-harvesting equipment are typical on many small- to medium-sized dairy farms. Investigate rental equipment options before investing in a lot of farm machinery.

any equipment you purchase. Farm machinery can be dangerous equipment to use if you don't have the knowledge or ability required to handle it properly. If you cannot find farm safety classes in your area, one alternative is to contract others to do the field and crop work for you. This dramatically reduces the amount of equipment you need to purchase, as well as your exposure to heavy machinery.

Remember, for the majority of the year, your equipment will sit idle. If you buy the equipment, you will be paying on an asset that is not working. Contract hiring of your crop work can help you get established without making the huge investment in machinery. Equipment expenditures can be substantial when getting started, and the less money spent on machinery the more that will be available for running your business. Generally, the local county agricultural extension office can help you locate the names of independent or custom operators in your area and provide typical custom rates.

Hiring someone else to do the field work not only frees up capital but also relieves you of equipment repair costs. Generally custom operators can accomplish more fieldwork in a shorter amount of time than you could on

Skid-steer loaders, or skidloaders, are reliable machines that can perform a variety of tasks. Their versatility allows you to lift, pull, push, and carry loads with ease.

Some beef-finishing facilities have drive-through alleys to disperse feed easily into fence-line feeding troughs. Intensive feeding requires daily observations of animals to avoid potential problems, such as bloat.

your own. It is always a good idea to talk over the terms of a contract with the custom operator and develop a plan for the timely harvesting of your crops during the year.

If your farm is not large enough for custom fieldwork, renting equipment may be a more attractive option than purchasing. Renting allows you to use appropriate equipment for the crop and then return it without outright purchase. Check with your local equipment dealers for information about renting equipment.

Many small items, including water tanks, hay feeders, forks, shovels, pails, wire, tools, gates, and halters for tie-ups, are necessary to run a farm. An easy way to learn what may be needed on your farm is to attend a local farm auction. The small items are typically sold at the beginning of the auction prior to the sale of larger equipment such as machinery and tractors. You may not need all of the items you see being sold but it doesn't cost you anything to attend the auction and get ideas.

CHAPTER 4

FACILITIES FOR BEEF— WHAT IS NEEDED, WHAT IS NOT

The type of facilities or buildings on your farm may determine the size of your enterprise and the number of animals with which you can easily work. In today's world, raising animals in the most humane way possible is of paramount importance and the facilities you have will help accomplish this.

INVENTORY BASIC FACILITIES

Your first step is to take an inventory of buildings available to you for raising cattle. The basic facilities include shelters, corrals, and feed storage areas. In many cases the buildings you already have will be sufficient in providing shelter for your animals. These buildings do not need to be elaborate or expensive, but they need to be sturdy enough to withstand the movement of cattle and the pressures that their bodies can exert upon them.

In most climates, some form of shelter is needed for cattle, whether it is to allow them relief from the heat, the cold, or something in between. Shelters do not need to be complex structures because beef cattle can acclimate themselves to most weather extremes if they have access to water and feed and if they can get out of the wind, rain, or sun.

A sturdy shelter will allow your cattle to get out of extreme weather conditions. Beef cattle do not need to be tied up so free access to shelter will be sufficient.

Windbreaks can be used around feeding areas to make conditions more comfortable in extreme cold for both owner and animals. Metal windbreaks can be constructed quickly and are solid enough to withstand most high-velocity wind speeds. Tree windbreaks can be planted on the outside of any corral or cattle area but it may take several years before they are effective. The location of your windbreaks may determine which kind you construct.

Properly constructed corrals will provide many years of trouble-free use in moving, sorting, loading, and restraining your cattle. The ideal corral provides safety for the animals and handlers to minimize the risk of injury. It also provides a path to gently and calmly move animals in pleasant working conditions.

A squeeze chute provides a safe, humane method of restraining animals. These heavy metal chutes allow an owner or veterinarian access to work with the cattle for vaccinations, tagging, branding, pregnancy checking, castrations, and many other situations.

If there isn't a building available that is sturdy enough to house your cattle, shelter construction needs to be part of your plans. Or you can consider converting an existing building not necessarily made for housing into a shelter.

Being able to easily and safely move your cattle into, out of, and through your buildings, lots, and corrals needs to be the top priority The more crowded and excited your cattle get from being moved, the greater chance for injury and bruising of their carcasses. Incorporating working facilities that reduce stress on cattle will help you handle the animals in a safe, calm, and humane manner when moving them through a chute for vaccinations, AI (artificial insemination), or for veterinary assistance.

It will be less work for you if the sheds, barns, other outbuildings, and corrals are easy to clean, refresh with bedding, and move around in with equipment. It may be worthwhile to renovate existing structures to accommodate some of these chores.

A fence-line feeding area provides an easy way to feed small or large numbers of cattle in a short period of time. An open-face fence-line feeding structure such as this allows you to place hay, silage, or grain where it can be eaten without being trampled.

A fence-line feeding system can be constructed with tie-ups that can be used as a replacement for a squeeze chute. These tie-ups are self-locking, which means they can be set so that when an animal puts its head through to eat, the top latch of the stanchion will close and lock the animal in place.

For a small herd, a tie-up can be installed in an existing barn. By encouraging cattle to use and have free access to the tie-ups prior to needing them, you will save time and frustration because they are already used to the tie-ups.

The health of beef animals is improved by keeping them in the natural air and not confining them inside a barn or shed. A lot of money is wasted in building expensive barns and sheds for cattle when thousands of animals are successfully raised each year with little or no shelter.

PROVIDING SHELTER AND WINDBREAKS

Windbreaks can play an important role in protecting your cattle, particularly young animals. Windbreaks work well in many cold climates because they protect animals from strong winds and the bitter cold. If the windbreaks are properly placed, they can provide protection to feeding areas, pastures, and calving areas. Reducing the wind speed during the winter lowers animal stress, increases feed efficiency, assists animal health, and makes the working environment for you a bit more pleasant.

The windbreak should be designed to meet the specific needs of your farm. There are two major types of windbreak designs: the traditional multi-row design and a newer twin-row, high-density design. The

Access to water is more important than access to feed. Providing fresh, clean water for your cattle helps them better utilize the forages and feeds you provide. Cattle with access to water will grow better and faster.

Beef cattle and dairy heifers can withstand inclement weather conditions if they have access to shelter, feed, and water. Extreme cold weather requires more volume of feedstuffs because of increased nutritional requirements.

design you choose depends on the area where you place it, the space available, and what you want the windbreak to accomplish.

The traditional windbreak can be made of three or more rows of trees, shrubs, or tall poles placed in the ground with galvanized sheeting nailed between them. The twin-row, high-density design is more elaborate and utilizes closer spacing between the rows and the trees and shrubs in each row.

While trees provide a good windbreak, they may take years of growth to reach maximum effectiveness. An alternative to trees is to construct a wooden or metal windbreak that can accomplish the same purpose. Designs for different types of windbreaks are available through your county agricultural extension office.

Corrals can be built with long, smooth wire panels attached to square wood posts. When covered with 2x4s, the panels provide a simple, sturdy fence line that will not injure animals that brush against it.

SETTING UP CORRALS

A corral's main purpose is to constrain animals for individual sorting, loading, and movement to other locations, or to temporarily hold them for feeding during the winter or when summer pastures need a rest from being grazed.

The proper construction of a corral can provide many benefits, including ease of moving cattle by removing the fear factor they may experience during sorting. By funneling them into a smaller space, the animals will feel less threatened than if they are herded wildly across the lot.

Cattle tend to follow one another. The movement of one cow or steer toward an enclosure will affect the behavior of the other animals. If the movement is easy and willful, the problems will decrease. Plan your corral so that you can move the animals with a minimum of excitement and stress.

CONSTRUCTING CATTLE RESTRAINTS

There will be times when you need to restrain an animal. This may be for artificial insemination, pregnancy examinations, or other health issues. Restraining large animals is necessary for your safety and those who work with you, as well as for the animal's safety.

Corner posts made from hollow steel piping that is cemented into the ground make excellent anchors for the fence and provide a solid base from which to suspend heavy metal gates.

Commercially manufactured squeeze chutes offer excellent restraint but they can be expensive. Plans are available from university agricultural extension services for building homemade chutes constructed of metal or lumber. These are generally adequate when handling a small number of animals.

Your barn may have an area that can be converted to a holding pen. With a little planning and reconstruction, a self-locking stanchion can be installed that will make it easier for you to catch the animal. You may decide to have two sizes, one for large animals and one for smaller animals, to be used for vaccinations and castrations.

PROVIDING FENCE-LINE FEEDERS

If you are planning on some form of feedlot for raising cattle during part of the year, you may find a fence-line feeding system beneficial. This is where part of the fence line is used as an area to feed the animals. Some type of bunk system is used to hold the feed and a cable or metal bar running along the top serves as a restraint so the animals

Open shelters with dry bedding materials make excellent housing for steers, heifers, beef cows, and dairy cows in their dry stage of lactation. These structures can be easily cleaned with loaders to save labor and time.

Confinement corrals can be constructed of heavy angle iron welded together. This will provide support and strength to last a long time with little maintenance, as well as safety from injury.

The loading chute can be designed so cattle cannot see what is ahead of them. Curves and angles in corral patterns keep cattle calm and help them move ahead inside the alleys.

Wide, sturdy gates allow trailers easy access into the corral for loading and unloading cattle, cleaning up with tractor loaders, and sorting animals.

Cattle-loading chutes should be constructed of heavy lumber. The chutes make loading and unloading easy with less risk of injury to the animals or their handlers.

Used materials, such as railroad ties and metal guard rails, are useful materials to construct a sturdy corral. Such a corral will last many years with little maintenance.

A well-planned, sturdy corral layout includes places to hold cattle while slowly working them through gates. Close confinement requires facilities that provide ease and safety for both cattle and handlers.

don't climb over the bunk. It is best to put down gravel or cement at the base of the feeding area for the animals to stand on while feeding. This will reduce the mud along the feeding area and make it easier to clean.

ACCESS TO WATER

Your cows most basic need is access to fresh, clean water. Typically, a beef cow will drink 12 gallons or more of water per day. Full access to water will help the cows withstand extreme temperatures of hot or cold better than if their water supply is limited.

Cattle require as much water during extreme cold periods as they do during hot temperatures. Contrary to popular belief, cattle cannot acquire enough water from snow to account for their daily needs. They use more energy and liquids to transform the snow into water than what they gain from it. The result is that cattle dependent on snow for their water supply become dehydrated and will die quicker from lack of water during a snowstorm than from a lack of feed.

A tank or some other structure holding a large volume of water can be placed in the lot where the cattle congregate. The size of the tank will be determined by the number of animals you have. Pasture water tanks come in various sizes and can be moved from pasture to pasture. A water line supplying these tanks can be laid on the ground or dug into the ground depending on your preference, the climate, and the terrain of your farm.

In uneven fields a large water tank can be set at a high point and used with a gravity flow system to fill smaller paddock tanks at lower elevations. Filling one large tank reduces the need to fill several smaller ones.

CHAPTER 5

ENCLOSURES— GOOD FENCES MAKE FOR A GOOD NEIGHBOR

A sturdy perimeter fence will keep your cattle from straying into your neighbor's fields and causing crop damage, or wandering onto highways and causing accidents. Perimeter fences are considered permanent. A well-constructed one will last for 25 to 50 years.

For generations, fences have marked the boundaries of farms and given the owners containment for their livestock and protection from animals on adjoining farms. Good fences are of high importance when raising cattle, not only as a mark of property lines but also as a way to humanely enclose animals. In many respects a fence will keep your animals in your pastures and your neighbors' animals out.

WALKING THE FENCES

When buying a farm, ask to walk and view the fence lines with the owner or real estate agent to fully understand the boundaries. Walking the fence rows will also give you a valuable perspective of the farm beyond looking at a plat map, survey map, or photographs of the property. Walking the fences will also give you a sense of proportion and provide you an opportunity to inspect their condition and make note of problem areas that need attention.

During recent decades, many old fence lines have been removed in order to till the fence rows to keep weeds, small shrubs, and trees from growing. If fence lines are missing, it is important to know where they belong before building a new fence. Tearing down an improperly placed fence can be costly, time-consuming, and frustrating. To be sure a fence is constructed along the proper boundary line, you may need to hire a surveyor.

Once you are on the farm, it is helpful to study an aerial map to get another perspective. Aerial photographs of your farm can be obtained from your county Farm Service Agency (FSA). Most, if not all, farming areas have aerial maps available and you can request a copy of yours. With their assistance, the relative boundaries of the fence lines can be seen. This map will show the farm's perimeter and give some indication of the lay of the land, which may be useful in handling any hills and valleys in the fence line.

An aerial photograph is also used to determine which areas of a farm are suited for different purposes. With the help of the FSA office, you will be able to identify those areas suited for hay or permanent pasture, woodlands that may or may not be pastured, and areas suitable for cultivated crop production. These land assignments are made based on uses that will best protect the soil structure and return the greatest possible profit for each type of soil.

Right: **A temporary fence requires less-sturdy materials than a permanent fence. Temporary fences may be easily and quickly moved to different areas of the field and expanded or contracted as needed.**

Below: **A woven-wire fence consists of several horizontal lines of smooth wire held apart by vertical wires called stays. These fences are strong but generally cost more than other options. Woven-wire fences last many years when properly constructed.**

A FENCING PLAN

Raising cattle on your farm will most likely require two different kinds of fences: permanent and temporary. Permanent fences are intended to last for many years with minimal repairs and should be constructed with sturdy materials. These are typically perimeter fences or fences around streams or waterways.

Temporary fences are intended to last only a short time. They do not need to be stoutly constructed and can be made from less-expensive materials. Temporary fences are usually found within the perimeter fences.

Certain areas of your farm may be more suited for pastures than crops but are located a long distance from your buildings. In this case, you may need to construct one or more lanes, or pathways, for your cattle to gain access to those areas.

The key to a successfully built fence starts with a good plan on paper because alterations can be made quickly if you change parts of your plan. This allows you to calculate the length of the fence and the amount of fencing materials needed before driving in a post. It will also help you estimate the costs of your fencing project.

Once your plan is developed on paper, walk out to your fields and see if your ideas will work or if you need to make changes before you start. It will probably be less expensive for you to construct your own fences rather than hiring out the work. Although you can hire contractors who will finish the project quickly (but at a greater cost), with a little experience and guidance you can build a fence that will withstand the pressures placed on them by your cattle and last a long time.

A key rule is to build straight fence lines wherever possible. They are easier to construct, retain their tension for a longer period of time, and require fewer materials. You may need to make curved fences in certain areas, but try to avoid these whenever possible.

Barbed-wire fencing is an effective and less expensive material for constructing perimeter fences. A four-wire barbed-wire fence is sufficient for keeping cows and calves within your fields. A tightly constructed barbed-wire fence can provide trouble-free usage for many years. Don't use this type of material in corrals or close quarters because of the risk of injury to the animals or handlers who come in contact with the sharp barbs.

Several strands of high-tensile wire can be electrified to keep cattle within fenced areas. The wire is held in place with adapters that provide even spacing between the strands.

Cable-wire fencing is used in corrals and pen areas, and consists of smooth, steel wire cables stretched from one anchor post to another. A heavy spring maintains tightness and helps absorb any shock to the wires.

PERMANENT AND TEMPORARY FENCES

A permanent fence that surrounds your farm is one of the best ways to protect your livestock investment. Besides establishing a fixed property line, perimeter fences are the last line of defense if your animals happen to escape from their designated grazing areas or the small lots around your buildings.

If your perimeter fences are intact it is unlikely the animals will be able to invade your neighbor's property, thereby relieving you of possible financial liabilities due to the destruction of property or crops. You cannot decide what level of animal enterprise your neighbors may have, so maintaining good fences for yourself will give you the same protection to your property by keeping their animals out.

Perimeter fences also keep animals from wandering onto highways and roads and protect your livestock and the driving public from possible highway collisions. If it is not possible to construct a permanent fence around the entire perimeter of your farm, consider building those sections that will be most useful to your enterprise and plan to add the rest at a later date.

Permanent fences should be considered for those areas that will be used for pasture several years in succession. Farm ponds and other waterways should receive priority for permanent fences to control livestock access or to allow access only for drinking. In areas where travel through waterways is necessary, access can be restricted with a permanent fence. You may also consider permanent fencing for fields where cultivated crops are grown and your animals are allowed access after harvest to graze the residue.

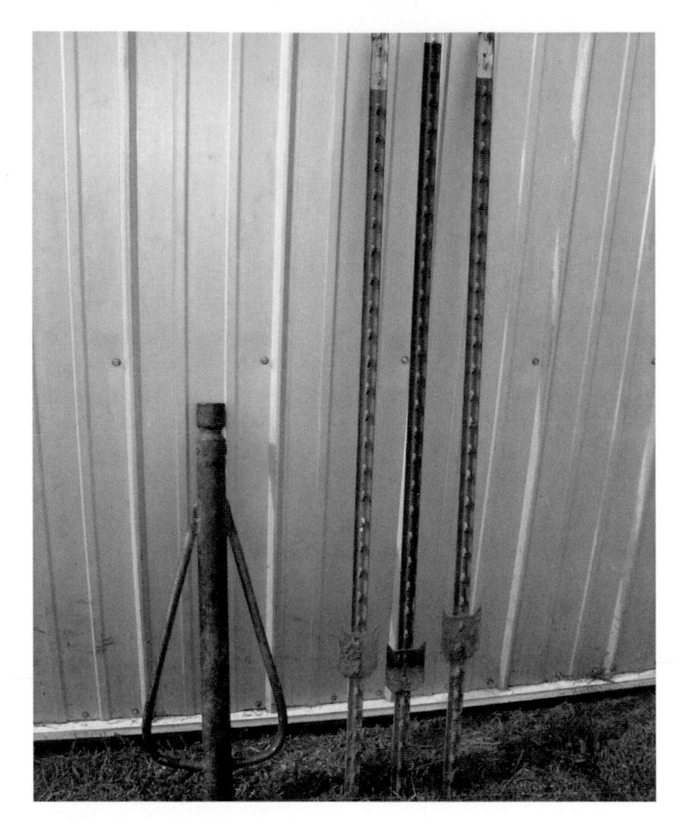

T-posts are the most common metal posts used in a fence line and come in several lengths. They are easily pounded into the ground by using a post hammer and can be removed by pulling them back out.

Temporary fences are used for a few days, weeks, or months and can then be removed. Movable fences are less expensive to build than permanent fences and are quicker to set up. However, they are less durable and may not be as effective in cattle control, particularly if maintained by an electric current.

One advantage of temporary fences is that they can be quickly and easily moved. By being able to relocate them every year or several times within one year, temporary fences offer flexibility for pasturing cattle by expanding or contracting the size of pasture. This flexibility allows you to accommodate any increase or decrease in the number of animals placed within a pasture, their relocation from pasture to pasture, and allows for pasture rotation.

FENCING MATERIALS AND VARIETIES

There are many types of materials that can be used to construct a permanent fence, but the most appropriate for cattle include barbed wire, woven wire, cable wire, high-tensile wire, and electric wire. Each of these materials may be a part of your permanent fence plans. Barbed-wire, woven-wire, and electric-wire fences will generally best suit your fencing plan. High-tensile-wire and cable-wire fences work well for confinement areas, such as corrals, around barnyards, and in loading areas.

A standard barbed-wire fence uses three to five strands of wire with posts usually spaced 10 to 12 feet apart. The more strands of wire in the fence, the less distance there is between them, which makes it more difficult for an animal to try to push its way through. Cattle will put any fence to a test out of plain curiosity or when there is a lush crop on the opposite side. A fence with five strands of wire will typically have spacing of about 10 inches between each strand and will be sufficient to deter cows and calves from trying to push their heads through.

Wood posts can be used at corners or placed in long fence lines as additional support for the wires. Wood posts can be round or square, depending on the source. They generally cost more than metal t-posts. Old wooden railroad ties can also be used as wood posts in fence lines.

Woven-wire fences consist of a number of horizontal lines of smooth wire held apart by vertical wires called stays. The height of most woven-wire fencing materials ranges from 26 to 48 inches. The height needed for your fence will depend on the size and athletic ability of your animals. If woven wire is used, it typically has a single barbed-wire strand running parallel along the top as an added deterrent.

Woven wire is often more expensive than other types of fencing material because of the additional metal used in making that type of fence. Woven wire can be two to three times the cost of barbed wire for a similar distance covered while having approximately the same life span. Also, woven wire is sold in rolls, which are more difficult to handle, and more roll units are needed to complete a similar distance than with barbed wire.

Cable-wire fencing consists of smooth, steel wire cables stretched from one anchor post to another. One end of each cable is attached to a heavy spring to absorb any shock on the wires, while the opposite ends are rigidly attached to the next anchor post. The wires may pass

The corner post is the most important anchor to building a good fence. Establishing the position of the corner post first will allow you to easily and accurately lay out the fence line. Using brace posts to support the corner post will ensure the corner post stays in place and keeps the wires tight.

Fencing can provide lanes to move cattle easily and safely from field to field. A lane built in a well-drained area will require less maintenance and keep cattle from creating muddy areas during wet weather.

Poly wire can be effectively used for temporary fencing needs. These plastic wires are embedded with thin metal strands and when electrified they will provide a small shock to any animal that touches the wire. Poly wire fencing is easily and quickly set up and taken down to give you much flexibility as a temporary fence.

through holes drilled in wooden support posts. Because of the higher cost of cable wire, they are mostly used for confinement areas and holding pens or corrals.

The purpose of a high-tensile fence is similar to the cable wire, although it consists of a single smooth wire that is held in tension between end-post assemblies. Five strands are often used with one or more of the strands being electrified to prevent animals from rubbing or scratching on them and moving them out of position. If stretched, these wires can sag and leave openings for the animals to push through. The advantages of high-tensile fence are that it leaves minimum damage to livestock hides, has a long life if properly maintained, and is easily adapted to specific needs.

Electric fences offer the flexibility of being used for both permanent and temporary fencing needs. Used as a movable or temporary fence, it can be made with one or two strands of smooth wire or a poly tape that has small wires woven into it. The poly tape is more flexible and easier to handle and move from one location to another. In either case, the wire is energized by an electric controller that receives its source from a standard farm electrical outlet

A cable-wire fence is typically anchored at two ends. In this instance, one corner serves as a load transfer instead of an anchor.

Plastic insulators attached to wood posts will keep the high-tensile wires separated and ensures an unbroken electrical circuit. These types of insulators are inexpensive and easy to install. They work well in all kinds of weather.

Corner posts may be anchored by pouring cement around the base. A metal corner post made from angle iron makes a solid anchor for stretching a four-wire fence.

or a solar-powered pack. Both types need to be grounded to complete the circuit.

While the advantages of an electric fence include easy movement from site to site and less bracing for corner or end posts, there are some disadvantages. The effectiveness of an electric fence diminishes when the electric current no longer runs the length of the wire or when vegetation grounds the wires. Livestock require training when first using electric fences and often will not be aware of the fence until encountering a shock from the wire. It's important the fence charger be maintained for full-time operation in order for the temporary electric wire to be effective in keeping cattle inside the fenced area.

FENCE MATERIALS AND CONSTRUCTION

The most important item you will need for fencing is a good pair of leather or heavy cloth gloves. These are absolutely essential when working with wire as they reduce the chance for injury and severe cuts to your hands if constructing barbed wire fencing. Some of the tools needed include fencing pliers, a posthole digger, protective eyewear, a tape measure or reel, and a wire puller. When you have determined the type and gauge of wire that will work best, you will also need staples, wood posts or steel t-posts, and clips.

Different classes of zinc coating on fencing wire have been established. Generally speaking, the higher the class

Lane gates to consecutive paddocks can be set with an electric wire and handle. Small lane openings can become muddy in wet weather because of the high volume of traffic through a narrow area.

number, the greater the thickness of the zinc coating on the wire, which leads to a longer life. Dealers in wire fencing can offer advice on the type and gauge of wire that will best suit your needs.

Fence posts are used to hold the wires apart and in place and are commonly made of wood or steel, each having advantages and disadvantages. Wood posts have an advantage over steel posts in strength and resistance to bending. Permanent fences often require decay-resistant fence posts. Wood posts that have been pressure treated can last as long as steel posts. Wood posts come in varying sizes so it is important that you use larger-sized posts for the corners and braces, while smaller-sized posts can be used as line posts.

Wood posts need to be long enough to support the fence height and the depth they are placed into the ground. A satisfactory post length is the height of the top wire above the ground, the depth of the post in the ground, plus 6 inches.

Steel posts have several advantages in that they cost less than a similar wood post, can be driven into the ground more easily, and they weigh less for easier handling. Steel posts generally are from 5 to 6 feet in length.

One main electric source provides the energy needed to electrify the high-tensile-wire fence. With the aid of strand connectors, a complete circuit can be maintained.

Most corner bracing is strengthened by using a diagonal crosswire. These can be used with a ratchet wire wrench to maintain a solid structure.

Fencing materials may be purchased at local farm stores, most lumber yards, and many hardware stores. Purchasing in bulk can save money. The approximate amounts of the various fencing materials required can be determined by calculating the distances involved. Use a standard roll-out tape measure or a reel tape, which measures the distance as you roll it over the area you walk.

The main ingredients to building a permanent fence with a long life span are solid end or corner posts, tight wires, and use of the right materials. Every fencing job has different requirements and each fence presents a slightly different approach than another. Like other construction and maintenance jobs around the farm, building a good fence requires proper techniques and a common-sense approach.

Building a new fence is easier if you first clear away old materials or brush and shrubs that may have grown in the fence line. This will leave you a clear path in which to work. If this is not possible, you should at least make the fence line clear from obstacles.

By locating the corner or end posts first, you will easily plan your fence layout because you will be aiming for those corners with your rolls of wire. At the same time you identify these end points, you will need to determine where the gates or permanent cattle lanes should be located.

You can build a good fence by understanding the importance of corner posts and the tightness of wire. To begin, lay out the fence line by first locating the corners, and gate openings. Set the corner posts where you make the first pull to stretch the wire. The posts and braces are similar in importance to the foundation stone of a building. Having solid corner and brace posts will make the initial

Shade trees in pastures provide an area where cattle can find relief during hot summer weather. If trees are available in pasture areas, consider constructing your fences to utilize their benefits wherever possible.

Breaker switches can be used to quickly and easily electrify different fences in a paddock system. Running from the main line, subsidiary lines can be turned on or off as needed.

tightening of the wires easier and will help hold the wire tension for many years.

Set the corner brace posts first, as this will give you a solid base to attach the wire puller to stretch the wire you have rolled out along the line. The corner posts can be set by driving them into the ground with a mechanical post driver, digging a hole and tamping them in with dirt, or cementing them into the ground.

A depth of 3 1/2 to 4 feet is recommended for corner and brace posts because of the tension exerted by the stretched wire. The shallower the depth, the more likely the corner post will eventually work its way out of the ground. A post moving vertically, even 2 inches, will provide enough decrease in wire tension that the fence will become slack and eventually provide gaps in the wire, thereby allowing your cattle an opportunity to push through the fence.

With sufficient depth, the post will not be affected by frost upheaval in colder climates. Depending on the soil type, the depth of the post may need to be increased, as well as anchors put into the bottom of the post to keep it from slowly working its way to the surface. Rising posts usually occur in mucky or marshy soil types where the ground surface is not as solid or stable as in other areas of the pasture. While it may take more time to anchor these kinds of posts, the result will be worth the extra effort.

After setting the corner post, the brace posts can be put in. The brace post provides support to the corner post because it transfers the pulling force of the wire from the end to the brace. This is important because when a wire is first stretched, the pulling force on the corner post may reach 3,000 pounds. Winter cold can cause contraction of the wire, which can increase the pull to 4,500 pounds. The corner and end post assemblies must be strong enough to withstand these forces or the posts will slowly be pulled out of the ground enough so that the wire loses tension.

You can set the brace posts by using the guide wire previously laid out to line them up or it may be just as easy to visually sight them. The distance between the brace and the corner post should be 2 1/2 times the height of the fence for maximum support.

There are several ways of setting line posts in the fence. Depending on the length of your fence, the spacing between wood posts is normally 20 to 25 feet, and the

Corner posts do not need to form 90-degree angles every time. Depending on the field configuration, the angle may be smaller than 90 degrees. Whether at right angles or not, corner posts will provide an anchor for the entire fence.

Without perimeter fencing, cattle will have access to roadways or your neighbors' fields. It also means that your neighbors' cattle will have access to your fields.

spacing between steel posts is normally 5 to 8 feet. If the total length of the fence is greater than 650 feet between corner posts, it is advisable to insert a braced line post every 650 feet. This will help strengthen the wire and maintain its tension over long distances. Braced line posts are also very useful when the fence is made over rolling land or hills. Once the layout for the perimeter fence is finished, you can plan for any interior fences, corrals, waterway barriers, or other enclosures.

Gates and passageways for your cattle should be located in the corner of the fields nearest to the farm buildings. Placing them in corners makes it easier to move cattle from one field to another and allows you to have your corner posts and gates in areas that do not break up the fence line.

A lane should be located in the driest areas possible, such as along a natural ridge or some other higher land feature. Cattle movement within lanes can quickly develop gullies from repeated use. If well-drained areas are not possible, consider using a lane fence that can be moved after several years or plan on covering the lane with gravel for better footing for your cattle. If possible, avoid placing lanes through wet or low-lying areas because the cattle traffic will soon turn that area into deep mud.

QUESTIONS ABOUT FENCING

Booklets explaining different fence constructions are available from county agricultural extension offices or fence manufacturing companies. Although there are initial costs involved in constructing a fence, there may be programs available through land conservation agencies that can help offset some of these costs when applied to certain management practices. Check with your county agricultural extension office to learn more about these programs.

When cattle first enter a pasture they will walk the perimeter fence to acquaint themselves with their boundaries. Well-built fences around your farm will eliminate escapes and the frustration in having to retrieve wandering cattle.

Above: High-tensile wire can be kept tight with a clamp-and-spring assembly. Each wire can be tightened manually as needed. The springs allow expansion and contraction of the wire without breakage.

A fence built in a straight line helps maintain tightness in the wire. Use wood posts and bracing for making curves and corners in the fence line.

CHAPTER 6

CATTLE TYPES— CHOOSING A BEEF BREED

Heritage breeds were developed in specific areas because they could adapt to and thrive in local environments better than other animals or breeds. This adaptability made them an ideal breed for farmers in a specific area. Their uniqueness has attracted contemporary cattle producers seeking to incorporate their desirable traits into their herds.

Before deciding which breed to raise on your farm, consider your preferences. Is a pedigreed herd important to you? Would you like to work with a heritage breed? Are you interested in showing cattle? Do you like a certain color of animal?

There are at least 60 beef breeds present in the United States. Researching the characteristics of each breed will help you select which breed or breeds you want to raise. Some breeds may help you create a niche market. This can draw interested cattle buyers to your farm, or you simply may have a lot of pleasure raising them.

The facilities you have may influence breed selection. For example, Scottish Highland cattle have long, sharp horns. This would need to be considered in transport, enclosures, and restraint methods. Don't be discouraged from seeking such breeds. Just understand breed characteristics and decide which breed best suits your farm.

The emergence of artisan food markets has increased producer interest in supplying those markets with animals from less common breeds. Interest spurred by the know-your-source food ethic has brought back old-time flavors and textures to the dinner table. The meat from these cattle allows restaurants to feature unique dining offerings on their menus.

Some breeds veered toward extinction before they were raised and promoted as heirloom breeds. This has led to an animal conservationist movement, which includes sustainable food groups and like-minded chefs, as breed-specific heritage meats became available to the artisan food chain. By working with a heritage breed, you will join with others who wish to restore vanishing breeds and the

Scottish Highland cattle are an example of a heritage breed. By working with heritage cattle, you join others wishing to preserve the nation's gene pool of unique breeds.

nation's livestock gene pool, which has been dramatically reduced by the breeding of animals for the mass market. Many heritage breeds thrive on grass-based farms because that was generally the only feed available as they developed into a recognizable breed.

As cattle began to be bred for commercial efficiency, they were fed a rich diet of grain to fatten them to market weight in a shorter period of time, usually under a year. It takes longer to raise animals on grass pasture, but forage intake allows cattle to turn grasses and legumes into muscle. Grass-fed cattle develop muscle rather than turning the energy they acquire from grass into fat.

The Scottish Highland breed illustrates this point. These cattle evolved in cold, rocky areas of northern Scotland and became hardy eaters that rarely became sick. Today, they can fatten naturally to 1,300 pounds solely on grasses and legumes.

Whether you decide to raise a more unique breed or acquire one considered more popular or mainstream, there are differences to consider. Knowing these differences will help give you more confidence in your choice.

A mixed commercial beef herd can weave many good traits of several breeds into your breeding program. By understanding the characteristics each breed has to offer, you can make the best decisions for your situation.

BREEDS FOR YOUR FARM SITUATION

Each farm is unique. Breeds best-suited for your purposes may be totally different from that of your neighbor. The region of the country where you live may influence your breed choice. Some cattle breeds are able to tolerate different climates better than others. Besides climate considerations, breeds also differ in traits, such as mature cow size and calf growth rate. This impacts nutritional requirements and eventual production costs. Breeds also differ in growth rate, reproductive efficiency, and maternal ability. Choose a breed that matches your goals for these characteristics.

Some breeds offer registry programs. This involves raising pedigreed animals or purebreds versus unpedigreed or commercial animals. The parentage of purebred animals can be substantiated and this may allow them to participate in purebred programs, such as shows and fairs. This may also open up a market for seedstock purposes.

The purebred versus commercial markets do not need to be competitive as they only differ in approach and end goals. Raising quality cattle can be accomplished whether they have a documented ancestry or not.

BREED VARIATIONS

Cattle breed associations can provide information to help you select the breed appropriate for your farm. Although individual animals may differ, some general guidelines and variations apply to each specific breed.

Typically beef breeds that produce heavier calves at birth also tend to have the heaviest calves at weaning—the time the calf is taken off milk from its mother and switched to feedstuffs. Those calves heaviest at weaning also tend to grow the fastest in a feedlot or pasture and reach market weight the quickest.

Breeds with heavier birth weights may have more difficulty at calving time and require more assisted births. Calf survival tends to be higher for breeds requiring less assistance at birth. Heifer calves sired by breeds with smaller mature size tend to reach puberty at a younger age than those that reach larger mature sizes.

Aberdeen-Angus is one of several British breeds imported into the United States in the nineteenth century. Black Angus have remained a popular beef breed because of calving ease and carcass quality.

CROSSBREEDING

Crossbreeding involves mating two breeds to produce offspring that potentially carry the best traits of each breed. It can have a positive influence on economically important traits, while reducing costs of production. Two advantages typically result from crossing two different breeds rather than using only one breed in a program: heterosis and the blending of the best qualities of each breed used.

Heterosis, or hybrid vigor, is the measurable gain resulting from combining the genes of two different breeds over straight-bred parents. The heterosis effect can be calculated by formulas that give weight to different traits for which heritability can be determined or studied. Some

Shorthorns make excellent grazers with good weight gains on pasture. They are adaptable to many situations and work well in crossbreeding programs.

Large-framed cows have the advantage of producing large calves more easily than small-framed cows. Large calves reach weaning and market weights quicker than small birth weight calves. One disadvantage of large-framed cows may be their increased nutritional requirements to maintain body size.

If left alone, the horns on this calf will continue to grow and eventually reach the approximate size of her mother's. Horns, although attractive, are a nuisance in most cattle-growing situations, as well as a danger to you, your family, and the other cattle on your farm.

Calves can be genetically dehorned through a breeding program that uses bulls heterozygous for the polled gene. This will eliminate the need for different methods of manual dehorning. Genetic dehorning is the most humane method available.

traits are extremely difficult to measure, such as disposition, and may have as much to do with the environment where animals are reared as the breeding impact.

However, generally the hybrid vigor effect can help improve traits, such as growth rate and fertility. Essentially any trait that falls under genetic control can be modified through crossbreeding. Some breeders use two breeds and cross back and forth in each successive generation where others may use several breeds in a rotation. For example, a farmer may use Red Angus and Shorthorn for the first generation offspring and then breed all the resulting females to a Brahman for the second generation. That generation of females is then bred back to a Red Angus, and so on. There are many different combinations and any successful crossbreeder will keep extensive records to track progress and identify trends in his or her breeding program.

POLLED VERSUS NON-POLLED

Some breeds have horns and some do not. This is determined by a simple genetic recessive trait. Those animals polled (naturally dehorned) or non-polled (with horns) require different management. Horns on animals pose a danger to handlers, other animals in pastures or feedlots, and other cattle during transport.

If you need to dehorn cattle, there are several methods available. Electric dehorners are used on calves up to two or three months of age when the horns grow too large for this method. Burning the horn buds kills the blood vessels feeding the horn and the horns eventually drop off as the head heals.

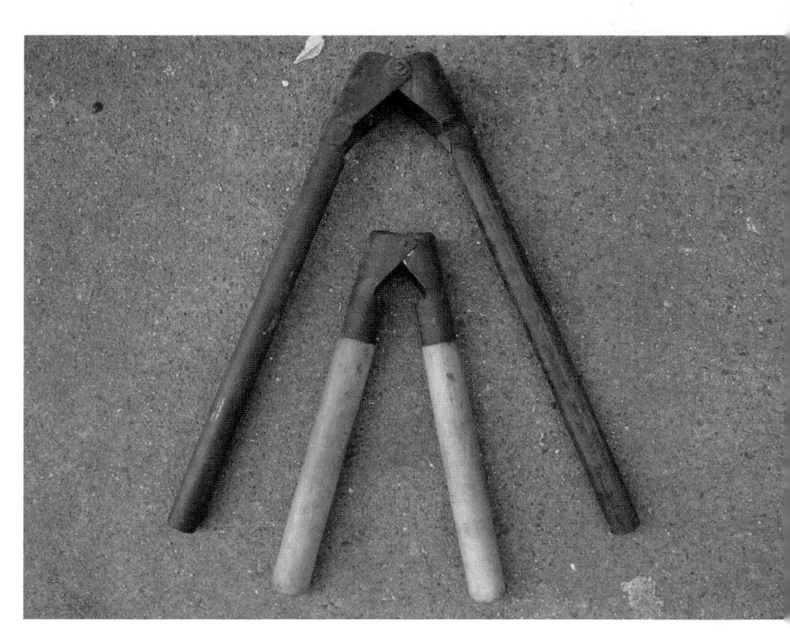

Gougers can be used to dehorn calves when the horns are too large for an electric burner. This involves using one of several types shown here, depending on the size of the horns. While this is an effective way to dehorn calves, it is time consuming and messy, and the animals need to be tightly restrained to do a proper job.

While aesthetically pleasant to look at, horns have no practical purpose in livestock today except in heirloom or heritage breeds to retain their original nature. Horns present a danger to people and other cattle.

Dehorning beef cattle through genetics is a more humane way of removing horns than by cutting, burning, or tipping them because there is no pain involved in the genetic method. Numerous studies are available that outline the economic costs of removing horns including labor, infections, and reduced growth rates while the wounds are healing.

The logical alternative to dehorning is by carefully selecting a breed that is polled or using polled bulls to sire calves that do not need dehorning. The polled gene is dominant and the horned gene is recessive. That means in one breeding season, a producer can take a herd of horned cows and breed them to a polled bull (one that is homozygous for the polled gene), and produce an entire calf crop of polled calves.

Occasionally polled animals will have scurs, small hornlike growths that appear in the areas of the head where horns would normally grow. Scurs are not connected to the skull by bony tissue so they are generally loose and moveable. Scurs can only occur on cattle that would otherwise be polled because horn growth overrides scur genes.

If you'd like to purchase polled cattle, ask if you are unsure about recognizing the polled characteristics. If you do not get a satisfactory answer, look at the head of the animal for signs of whether it is naturally horned or not. In smooth polled cattle the peak of the poll often increases. Cattle with scurs generally have more rounded heads but with a tendency to peak at the poll. Horned cattle generally have flat or only slightly rounded heads.

The cow in the rear and the young heifer in the center and the steer on the left are naturally polled and have a narrow poll at the top of their heads. The steer on the right was horned and had the horns removed. As the horns grow and develop, the bone structure on the top part of the head widens as it accommodates this growth. These signs give an indication whether an animal, whatever age, is polled or was horned.

The following is a sample list of the more popular beef breeds in the United States.

The South Devon breed is an English native that exhibits a good growth rate, excellent carcass quality, and are naturally polled. Their red color closely matches that of the Red Angus but the two breeds are genetically far apart.

Aberdeen-Angus—Black hair coat. Polled. One of the most popular breeds in the United States. Calves are usually smaller at birth with less difficulty for the mother at calving time, thus less calf loss. Weaning weights are usually equal to other breeds. Medium in size. Well-muscled.

Brahman—Usually gray but can be red. Mostly drooping ears. Grunt instead of moo. Good longevity. Withstands heat well and bothered little by insects. Can graze on poor-quality forage.

Charolais—Widely used in crossbreeding programs (using a bull from one breed on a cow from another breed, or vice versa). Light cream in color. One of the largest beef breeds. Long-bodied. Good size. Heavily muscled.

Chianina—Originated in Italy. White with black switch on the tail. Most Chianina cattle in the United States are now black. Possibly the largest beef breed in the world. Noted for rapid growth. Fine-textured meat.

Devon—Oldest of all English breeds. Red in color, skin is yellow. Have horns curving upward that are creamy white with black tips. A dual-purpose breed with good milking ability and good quality beef.

Dexter—First imported to the United States in the early 1900s from Ireland. Its small size requires less pasture space. Beef from Dexter breeds result in small cuts of prime, dark meat. Thrive in hot and cold climates and do well outdoors year round. They are dual purpose, being raised for both milk and meat, making them the smallest of the dairy breeds.

Galloway—Polled. Short legs, blocky, and compact. Long, thick, curly hair that is black to silver with some browns. Native to Scotland and are good foragers, extremely hardy, and able to stand cold weather conditions. Smallest of beef breeds. Can utilize coarse grasses and other forages usually avoided by other breeds. Produce a delectable meat.

Hereford—Rich red body color with white face. Considered to be the first breed to be developed in England. Muscular with well-developed valuable cuts: the back, loin, and round or hind quarters. Vigorous with good foraging ability.

Limousin—France native. Red-gold color over the back that gradually lightens to yellowish straw color under the belly and around the legs and mouth. Most U.S. Limousin are now black in color. Heavily muscled.

Murray Grey—Native to Australia. Solid dark to silver grey in color. Rapid growth rate. Easy calving. Docile dispositions.

Polled Herefords—Similar to Herefords in most respects but preferred by those who dislike the horns. Red color with white face.

Red Angus—Attractive deep red color. General characteristics are similar to Aberdeen-Angus. Have the same ancestry except for the color. Easy calvers.

Red Poll—Developed as a dual-purpose breed. Small and hardy. A true red in color, they sometimes have a white mottled face. Produce an exquisite meat.

Santa Gertrudis—Developed in Texas from crossing Brahman beef-type bulls on beef-type Shorthorns. Make large gains on grass and can forage areas of sparse vegetation. Tolerate heat and insects.

Scottish Highland or Highland—Developed in Scotland. Color may be black, brindle, red, light red, silver, or yellow. Small in size but extremely hardy. Long, heavy outer hair coat and soft, thick undercoat. Can withstand severe weather conditions.

Shorthorn—Native to Northeastern England. First established beef breed in the United States. Color ranges from red to white to roan or spotted. Large, well-muscled. Liked for crossbreeding purposes.

Simmental—Originated in Switzerland. Color is light red or cream with white faces like Herefords. Many U.S. Simmental cattle are now black in color. Docile. Good milk production for young. Well-muscled. Popular for crossbreeding.

The two steers shown here represent a Red Angus-Milking Shorthorn cross. Cattle can be crossbred into many different combinations to increase hybrid vigor and produce quality animals for market.

The steer and heifer shown here are a result of a Red Angus-Milking Shorthorn-Hereford cross. Crossbred animals exhibit excellent weight gains on grass pastures or in confinement programs.

A bull pastured with cows and calves can be dangerous. Some breeds, such as Red Angus, have a more docile temperament and do not pose as much of a threat. However, be wary of any bull.

There was a time when farmers throughout the world raised many different cattle breeds. However, as the industrialization of farms and farming has evolved, methods have favored fewer breeds that exhibit certain genetic traits that adapt well to intensive farming practices.

Previously, heritage or heirloom animals were selected for traits that made them well-adapted to local environmental conditions. The loss of these unique breeds over time has meant the loss of the valuable genetic diversity they possessed. When this happens, their genes can't be used to breed new traits into existing livestock, such as the ability to withstand climate changes or resist disease.

A growing number of farmers are interested in preserving threatened breeds. Heritage breeds may not be as commercially productive as other breeds bred to produce lots of milk, gain weight quickly, or yield particular types of meat. However, they may be better adapted to withstand disease and harsh environmental conditions. By choosing to raise a heritage breed, you can contribute to sustaining a unique population of cattle for the future.

Naturally polled cows will always produce polled calves when bred to polled bulls. Dehorning calves through genetics is a humane approach to removing horns and involves the selection of polled bulls.

The American Livestock Breed Conservancy has identified several beef breeds as being threatened with extinction. These may be breeds to consider establishing on your farm.

British White/White Park—White with colored points such as ears, nose, eyes, teats, and feet but not the tail switch. Native to Ireland, Wales, and England. Came into the United States in the 1930s by way of Canada to the Bronx Zoo in New York. Medium-sized animal. High-quality meat. Great genetic distance from other beef breeds.

Florida Cracker/Pineywood—Florida's equivalent to the Texas Longhorn. Descended from cattle imported into the Americas by the Spanish. Smaller in body than the Texas Longhorn without extreme horn length. Can withstand heat, insects, and humidity better than most beef breeds. Well adapted to low-input beef production. Breed includes almost all of the solid colors and many of the spotting patterns known to cattle.

Kerry—Descendants of ancient Shorthorn cattle of Ireland. Found grazing marginal pastures. Globally

Cows recognize their offspring for a long time. The maternal instinct is a strong natural impulse that helps nurture the calf and provides protection when it is young.

rare. Few Kerrys exist in the United States and a few herds are located in Canada. Mostly black in color, cows weigh 750 to 1,000 pounds. Typically a dairy breed. Can be used for beef production as cows are hardy and long-lived; many are still calving at 14 to 15 years of age.

Randall/Randall Lineback—Purebred remnant of lineback-patterned cattle once common in New England. Multipurpose cattle used for beef, dairy, and oxen. Lineback characteristic. Blue-black color with white line sweeping down the back from the shoulder to the rump and tail. The roan coloring on their sides varies from black to nearly white. Medium sized. Good growth rates. Distinct from the American Lineback, which includes any dairy animal with the lineback pattern. True Randall Linebacks are rare.

South Devon—Red in color. Native to Cornwall and Devon, England. Came to the United States in 1969. Easy to handle. Docile. Excellent carcass quality. Good growth rates. Strong maternal breed; good mothers.

A crossbreeding program may include several breeds in different combinations. Keep good breeding and calving records to identify ancestry, as well as to help decide which breed to use in the next generation.

Small-framed cows can be very efficient feed converters because less feed will be needed for body maintenance. Avoid breeding small-framed cows to bulls that sire large-birth-weight calves.

Dairy beef can be raised from all dairy breeds. This steer represents a Jersey-Milking Shorthorn cross. Dairy steers can grow well on pastures or in a confinement program.

CHAPTER 7

BEEF PRODUCTION SYSTEMS— WHAT'S YOUR PLAN?

A mix of several grasses in your pastures will provide a more attractive diet for your cattle than single-variety fields. Grass-based diets are good for the digestive system of cattle.

Once you have decided on a breed of cattle, the next step is deciding what type of production system you will use. Depending on the size of your farm, facilities and forages available, and your inclinations, you can choose from a number of options for raising beef: You can raise your animals on a grass-based diet, where your pastures provide the largest portion of their nutrition; you can raise your animals in confinement areas near your buildings and bring the feed to them; or you can work with a combination of the two. Although other options are available, a few major ones will be detailed here. There are systems that use grain as a supplemental feeding program in conjunction with pasturing and that may be an approach you wish to consider and possibly incorporate into your plans. Your county agricultural extension agent can provide information on other beef production systems.

GRASS-BASED AND PASTURE-GRAZED

Cattle can be raised exclusively on a grass-based diet. This type of program is gaining much interest and notoriety because of the possible health benefits derived from the meat produced under these conditions.

The low-fat diet that has become the focus of many people in the United States can have an impact on your cattle-raising program. Many people have tried to cut fat from their diets and, in doing so, may have decreased the amount of conjugated linoleic acid (CLA) they consume as well. Ironically, CLA has been shown to be a component of fat that can slow the progress of some types of cancer and heart disease, as well as diabetes and obesity. CLA appears to actually help reduce body fat and increase lean muscle mass.

CLA is a fatty acid that occurs naturally in many foods and is especially high in milk and meat from ruminant animals because the acid is produced in the rumen. Recent field research by the University of Iowa has documented concentrations of CLA that are three- to four-fold greater in the milk from intensively pastured dairy cows. Rib eye steaks from cattle finished on a combination of pasture and concentrates (grains) were higher in CLA than steaks from cattle finished on stored forages and concentrates. The researchers concluded that pasture grazing is "an effective method to improve the healthfulness of milk and beef."

Their study also concluded that a likely explanation of why pasture-feeding increased CLA content of beef and milk relates to the fatty acid composition of the grasses. Linoleic acid accounts for the greatest proportion of the fatty acids in pasture grasses. This fatty acid can be degraded to vaccenic acid in the rumen with transvaccenic acid being a precursor for CLA synthesis in the udder and in meat.

Also, grass-fed beef contains more Omega-3 fatty acids than grain-fed beef. Omega-3 fatty acids have been linked to reduced incidence of mental disorders and heart problems in humans.

This is not an extensive discussion of all possible impacts a grass-based grazing program can have for you but it gives some indication of the possibilities that exist. These considerations may be part of your program in developing meat products through different growing options.

COMMERCIAL COW/CALF

Purchasing pregnant cows that calve soon after they arrive on your farm will give you a fast start. The cow and her calf can graze together and the calf will slowly switch from milk to grass as it grows older. Calves as young as two or

A cow-calf pair is at the heart of any pasture-grazed cattle program and represents nature's entire cycle, from birth to reproduction and everything in between.

Young calves may be purchased and easily put onto a grass pasture. They readily acclimate themselves to their new surroundings. It is easier to take calves off of concentrates or grains and put them on grass than the other way around.

three weeks old will start nibbling on grass as they follow their mothers across the pasture.

This option allows you to raise the calves for a season before selling them at weaning, as yearlings at 700 to 850 pounds, or as finished beef ready for market. You also have options of buying bred heifers, young cows, aged cows, or buying pairs of cows and calves together.

Typically, cows are calved either in the spring or fall. A spring calving schedule will allow you to take advantage of the increased forage and nutrient production from your pastures that can support more animals. By fall or early winter the calves will have reached a sufficient weight so that they can be sold as feeders, or you can continue to feed them through the winter. A fall calving schedule will allow you to raise the calves on milk during the winter and sell any cull cows (cows that can't be bred, are old, in poor health) in the spring, if needed, when market prices are typically higher.

CALVES: WEANING TO FIVE MONTHS OF AGE

Day-old calves may be purchased with the intention of growing them to weaning at four to five months of age. Because of the fast turnover for these young animals, this option will not require as much pasture space and it may be a way to get started on your farm if older animals are not available.

Planning and timing will be the most important considerations with this option. Your climate may determine when you purchase calves. When they are transported, you will need to be aware of the stress on a young calf from hauling. Young calves are susceptible to illness if transported in inclement weather so a suitable vehicle or open-air carrier is essential for hauling them.

Because these calves no longer nurse on their mothers, they need to be fed milk or a milk substitute because the rumen (stomach) of a young calf is not sufficiently developed

Raising dairy replacement heifers may require a long-term commitment and the use of grains and concentrates to reach target weights for acceptable growth and breeding purposes.

Raising dairy heifers is compatible with raising beef cattle. Both will make good use of pastures available and grow at comparable weights if enough forages are available.

to digest grain, grasses, or hay. These can slowly be added to a young calf's diet but during the first week of life milk is the most essential nutritional requirement.

Feeding very young calves requires setting a consistent feeding schedule. Obtain whole milk from a dairy or use a commercial milk replacement powder that is mixed with water to feed them.

Calves need to be fed twice daily and perhaps more frequently if the weather is very cold (or very hot) as they will use the energy from the milk to maintain body heat (or to stay hydrated) as well as for growth. A suitable confinement area in cold weather, free from drafts with dry bedding material and low humidity, is absolutely essential for growing healthy calves.

To be successful with calf raising, you will need to learn about buying and selling markets for calves if you do not plan to keep them past weaning.

WEANING TO YEARLINGS

If you don't wish to raise calves, you may want to consider purchasing weaned animals from a market or sales agent. These calves will weigh 300 to 450 pounds when purchased and you can raise them to yearlings weighing 700 to 800 pounds.

One advantage of raising weaned calves is they can be put onto pasture quickly. However, animals arriving on your farm need to have access to dry hay prior to being let out on pasture. This is done to fill their rumen (stomach) before eating lush grasses or hay, which might cause gastric problems if their rumens are empty upon arrival. Once they have eaten their fill of dry hay, they can be given access to pasture.

Because they are older and larger animals than calves, you will need handling facilities that will be strong enough to withstand the added pressures of larger body weights.

Other facilities needed, depending on the time of year the calves are acquired, include shelter from inclement weather such as ice and wind, and shade during the summer. Continual access to water, whether in a lot or pasture, is essential.

If you are not raising the animals past the yearling stage, you will need to locate a market at which to sell them or perhaps consider selling them privately.

Grazing begins in early spring when grass starts to grow. Stockpiling a pasture from the previous season will provide forage but it will not be as good as the quality of new growth.

YEARLING TO FINISHING

Raising animals from yearlings to a finishing weight of 1,200 to 1,300 pounds is typically done by feeding a high concentrate grain diet and generally does not involve pasture grazing. Raising this size of animal requires very sturdy facilities. Close confinement production also requires a plan for handling and hauling the large quantities of manure produced by these animals.

STOCKERS AND GRASS-BASED FINISHING

Another production option for buying cattle and finishing them on grain is to put 600- to 700-pound animals on pasture. These may be animals you have purchased or raised yourself.

You can start them in the spring, taking advantage of the summer and fall growing seasons to enhance weight gain. However, without the use of a supplemental high-concentrate grain diet, it typically takes longer for these animals to reach market weight, perhaps nine months or more.

This option requires matching the stocking rates to the grass available. Every farm has an optimum stocking rate in the number of animals that can be sustained on the acreage available. These calculations are included in Chapter 9.

Grass grows quickly in spring. Moving cattle quickly through an early pasture rotation will help control the lush first growth and set up your pastures for the season.

Dairy steers can be raised on pasture once they are old enough to digest grasses and legumes. Raise dairy steers to diversify your farm and boost income.

Whichever option you choose, pastured cattle must have access to water, shade during the summer heat, and shelter from inclement weather. Choose an option that fits your climate and work schedule and be sure that you have adequate pasture for the number of animals you plan to raise. Your success will be determined in part by learning about cattle markets and market trends so that you can achieve the best returns.

DAIRY BEEF

Producers raising dairy steers find the low initial investment in dairy bull calves to be an advantage over buying beef bull calves of the same age. Dairy steers are castrated bull calves born in any dairy herd and raised to a target weight. They may be incorporated into a program from weaning to five months, to yearlings, or to finishing.

On pastures, dairy beef have shown uniform weight gains when compared to beef breeds, especially in the spring with lush plant growth. However, dairy beef may be discounted in some markets because they typically yield a heavier carcass.

Raising dairy beef requires intensive management particularly when the calves are very young. Beef calves typically stay with their mothers for several months and may nurse as many as eight times a day, consuming many pounds of milk. Dairy calves are generally removed from their mothers soon after birth and will rely upon you to provide the milk or milk substitute several times a day in order to receive enough nutrition.

DAIRY REPLACEMENT HEIFERS

If you wish to raise cattle other than beef, you may choose to raise dairy heifers. The number of independent heifer growers is increasing. Many large dairy operations now contract the growing of their heifers away from their main farm. Partly due to labor costs, these dairies tend to concentrate their efforts on milk production rather than raising their calves as replacements.

In those cases, the calves are quickly moved from their facilities to the heifer grower's farm and the care then transfers to the grower from that point until the same animal becomes a springing (pregnant, close to calving) heifer. This entails roughly an 18-month commitment by both parties. If you are considering this option, a contract that identifies the responsibilities and expectations of each party is highly recommended.

Because of target growth weights at different age intervals, it may become necessary to provide grain supplements to the diet of these heifers.

As with the other options, housing facilities are necessary because of the length of time from arrival to departure for each animal. Animal identification and record keeping is absolutely essential especially if you have more than one owner's animals on your farm. You must be able to identify the owner of each animal to avoid problems. Forms of identification include ear tags, freeze branding, retinal scans, tattoos, photographs, neck chains and other types of identification routinely available from livestock supply companies.

Raising dairy heifers requires attention to any health problems, presence of internal and external parasites, and other conditions such as bloat. You must handle health problems immediately. In addition, being proficient at

Dairy steers may be raised in confinement in large or small groups. A confinement program typically uses grains and concentrates to finish steers for market weight.

Calves of similar size are easily housed together. Dehorning when calves are young avoids setbacks in growth rates. Appropriate vaccinations assure fewer health problems in the future.

A row of individual calf hutches makes for easy feeding. Pails set or hung along the outside panel encourage calves to eat away from their bedded area.

artificial insemination, animal health protocols, and daily heat detection for heifers between 13 and 16 months of age will be necessary.

If acquiring animals on a continuous basis, you will need a plan to separate smaller animals from larger, more mature heifers in order to avoid physical harm.

There are advantages to this cattle-raising option. There are no initial investment costs in the heifers and you can negotiate your agreement for monthly payments to secure a regular income. Satisfied customers are likely to stay with you year after year. Be aware that losing one animal or calf can make a big difference in your final returns. Therefore good husbandry and good animal health is essential for a successful program.

OTHER CONSIDERATIONS

Call livestock markets to find out information on the current market trends and prices before you buy or sell any animals. Ask about the best times in your area to buy and sell your animals. Most marketers have an interest in wanting you to succeed because your animals also impact their livelihood. The more animals they sell through their market the more income they receive on commissions. It is to their advantage to help you find the best market for your animals. With their network of contacts, they also will have knowledge of private individuals who may wish to buy your animals, depending on what age of animal you have to sell.

LIVESTOCK PREMISES REGISTRATION

Some states are starting to require premises identification. This can help the agriculture department veterinarians identify and locate the possible source of an animal disease outbreak, such as bovine spongiform encephalopathy (BSE), or foot and mouth disease. In some counties, you may register at your local Farm Service Agency (FSA) office or contact your county agricultural extension office for information on how to register your premises on the Internet. The registration process is easy and there is no cost to the producer to apply.

CHAPTER 8

BEEF PRODUCTION SYSTEMS—ORGANIC, SUSTAINABLE, AND CONVENTIONAL

Most animal-production agriculture falls into three categories: organic, sustainable, and conventional. The organic and sustainable systems are gaining popularity with farmers and others for several reasons. The most prevalent are the increased markets for products from these systems and the increased value those products bring to the marketplace. There are significant differences between organic and sustainable farming and conventional farming practices. Consider each carefully before deciding which one to pursue.

CONSUMER TASTES

Conventional farms use intensive planting and harvesting of crops—aided by crop chemicals and commercial fertilizers—to increase production and generate higher profits. Much of this has been driven by consumer preference for cheap food. Shifts in consumer tastes and a higher demand for foods that come from organic and sustainable farms are making these production systems attractive to farmers. This consumer shift, which cultivates the image of healthier foods, has brought more money to organic and sustainable markets. It has allowed farmers to utilize practices that may be more in tune with their personal ethics and nature's

Organic and sustainable production systems have become popular with both farmers and consumers. This approach is also attracting more farmers to grass-based beef- and dairy-raising systems.

Pasture-grazing programs are increasing in popularity as a low-cost production system. Farmers concerned about the long-term effects of chemical usage on the soils, water supplies, and wildlife have sought out alternative methods of farming.

Grass-based programs are beneficial to the land because the grasses are deeply rooted in the soil. Soil erosion is minimal where grass farming is used, especially when compared to row-crop farming.

harmony. Organic and sustainable farming can be a good fit for small-scale farming.

ORGANIC FARMING

Organic farming is a holistic, ecologically balanced approach to farming that supports, encourages, and sustains the natural processes of the soil and the animals. Those involved with organic farming have a perspective that reaches beyond the present and requires a long-term commitment to their program.

In many ways, organic farming has been a reaction to the industrialized and large-scale, chemically-based farm practices that have become the norm in food production since the end of World War II. Organic farming emphasizes management practices over volume practices; considers regional conditions that require systems to be adapted locally; and supports principles, such as environmental stewardship.

Farming organically is a method where the whole ecosystem of the farm is incorporated into the production of animals or crops. It seeks to obtain the greatest contributions from on-farm resources such as animal manures, composts, and green manures for soil fertility and to eliminate external additives, especially synthetic chemicals.

Organic farming is flexible enough so that if sufficient quantities of materials cannot be produced on the farm, then off-farm nutrients such as natural fertilizers, mineral powders, fortified composts, and plant meals from approved organic sources can be applied without risking certification.

Organic farming promotes the health of the soil by encouraging a diversity of microbes and bacterial activity. This in turn enhances the growth of plants and grasses deemed healthier for the grazing animal. There is a beautiful symmetry to this type of farming where feeding the soil feeds the plant that feeds the animal that feeds you.

ANIMALS ARE WHAT THEY EAT

The old saying that you are what you eat also holds true for livestock. Cattle fed grasses or other roughages grown without synthetic chemical assistance will not ingest chemicals into their systems. This supports the confidence of consumers that they will not ingest chemicals from the meat you have produced.

Animals fed a diet of roughages grown under conventional chemically-assisted farming have no choice but to eat plant material that absorbs chemicals applied to the soil. These chemicals then move through the animals' bodies and are deposited back on the ground. By eliminating chemical usage, organic farming is gaining in popularity because of the perception that it produces a more natural, healthier product.

Consumers want to believe that smaller herds are treated more humanely, are allowed to move about freely, and eat what nature intended them to eat—grass, not grain—and that these practices produce healthier meat or milk.

NATURE'S FRIEND

Organic farming protects the soil from erosion because of the crop rotation used and the grass-based feeding program for the animals. It also helps build-up the soil by promoting the use of a diversity of crops. Additionally, one often-overlooked benefit from organic farming is the increase in insect and bee assistance in pollination of the crops.

The real strength of the organic idea is its attempt to reverse the industrialization and standardization trends that have tried to transform farming into factories. Many people involved in organic farming believe animals, as living beings, cannot be molded into an assembly line production model and sustain it for a long period.

Many farms and feedlots are designed to operate as nearly as possible to biological assembly lines with raw materials coming in and finished products going out, with the farmer doing a certain amount of tasks in between to accomplish this transformation of feed to food. Organic

Traditional farming practices have been transforming to utilize renewable resources, such as grass pastures, to provide nutrition. Erosion is limited along hillsides and valleys when grass filters are left intact.

farming breaks this cycle because its proponents believe that organic is as much a philosophy of life as a physical characteristic of the foods they produce. "Produced in harmony with nature" is more than a slogan, it is a belief in the fundamental diversity of nature.

ORGANIC CERTIFICATION PROCESS

The term organic is defined by law in the United States and the commercial use of organic terminology is regulated by the government. Certification is required for producing an organic product that is packaged and labeled as such for sale. While the process to become certified can seem extensive, the result is that products produced under certification are authenticated.

Standards for organic certification are set by the government or various organizations in which farmers can become members. Lists of these organizations are available from most county agriculture extension offices or from many states' Departments of Agriculture.

If you choose to farm organically, you need to study your soil and have it analyzed to determine which nutrients need to be added to balance its components for optimum plant growth and nutrition. You will be feeding the soil so that it can feed the plants. Grains may be fed to your animals but, under organic rules, only from sources that produce it organically in order to maintain your certification. Specialized organic markets must be found to purchase feedstuffs used on your farm and for the sale of livestock to get the greatest benefit from your efforts.

If your farm has used conventional practices with chemicals, there is a three-year transition period before organic certification can be obtained. During this time no chemicals of any kind can be used on the land or on the animals, antibiotics cannot be used to treat the cattle, and genetically modified organisms (GMOs) cannot be planted.

OPPORTUNITIES WITH ORGANIC BEEF AND DAIRY PRODUCTION

The exclusion of insect- and pest-control products, as well as antibiotics for use on animals, can make raising cattle seem more of a challenge. But it is being done in many different parts of the country with great success because of a change in traditional thinking by those working with that

Watersheds are large areas that drain to a common waterway, such as a stream, lake, estuary, wetland, or ocean. Individual actions can directly affect water quality. Develop plans and uses for soils, nutrients, and water systems that preserve water quality.

type of farming. There are opportunities and it is possible to accomplish your goal of organic farming by studying what you put onto your land and what meat or milk you harvest from it for the marketplace.

SUSTAINABLE FARMING

Sustainable farming is a goal rather than a specific production system. Although they are often thought to be the same thing, sustainable and organic farming are not necessarily synonymous. The goal of sustainable farming is to approach a balance between what is taken out of the soil and what is returned to it without relying on outside products. The aim is for perpetual production.

The idea of sustainable farming starts with the individual farm and spreads to communities affected by this farming system, both locally and globally. Many people concerned with sustainable issues take a global approach with this system because it does little good for the overall goal if your farm has a negative impact on the environmental quality somewhere else. Sustainable farming requires that outside products applied to your land are available indefinitely and nonrenewable resources need to be avoided.

Growing and harvesting crops removes nutrients from the soil and without replenishment the land becomes less fertile. Therefore, one of the major keys to success in sustainable farming is soil management. Those following this practice believe that a healthy soil is a key component of sustainability. A healthy soil will produce a healthy crop that has the optimum vigor and is less susceptible to insects and pests.

If you are buying a farm or if you are already farming, it is likely you will need a transition period to become sustainable. This transition is a process that generally requires a series of small, realistic steps. Because of the costs involved, your family economics and personal goals may influence how fast or how far you will go in this transition.

Grazing programs provide an ideal opportunity for raising cattle for organic markets. Low production and maintenance costs, in addition to increased prices for organic meat and milk products, make this production system attractive.

CATTLE PRODUCTION PRACTICES

Raising cattle using sustainable production principles will include consideration of how you integrate your crop and animal systems. The key components are profitability, management of your animals, and stewardship of the natural resources, such as water and soil.

Proper grazing methods and pasture management can eliminate the most adverse environmental consequences associated with your animals, including stocking rates, so that limited amounts of feed must be brought in from the outside. The long-term carrying capacity of your farm must take into account short- and long-term dry spells and reduce overgrazing on fragile areas of the farm.

Animal health is an important part of a sustainable farm because the health of your animals greatly affects reproductive performance and weight gains of calves. Unhealthy livestock waste feed and require additional labor.

SOCIAL CONTEXT

Sustainable farming asserts that the stewardship of all natural and human resources are of prime importance. Stewardship of people includes consideration of social responsibilities, such as working and living conditions of farm laborers, the needs of the rural communities, and consumer health and safety both now and in the future.

Sustainable agriculture requires a commitment to changing public policies and social values, as well as preserving natural resources. This may include becoming involved with issues, such as food and agricultural policy, land usage, labor conditions, and the development and resurgence of rural communities.

Developing a strong consumer market for your products can help sustain your farm. Becoming part of a coalition can be useful in promoting your products and educating the public in general. By educating more of the public, you will likely increase your market and therefore continue the cycle you started.

Practicing and implementing sustainable farming can be a challenging but rewarding option. For those who believe in preserving the land and water for our use, as well as for future generations, this practice can have benefits far beyond any monetary compensation.

Organic and sustainable production systems make use of solid manures to provide nutrients for the soil. This helps decrease the need for outside fertilizer purchases.

SUSTAINABLE BEEF PRODUCTION

Sustainable beef production relies on pasture-based production methods that lower the costs of mechanical harvesting because the animals harvest the grass themselves. In one sense, a cattle producer is a grass farmer and the product you sell is grass, only it is sold through the cattle you raise.

One advantage of this system for raising beef is that lower costs are involved in raising the cattle. The elimination of expenses for chemical applications to control insects, pests, and weeds not using fertilizers greatly increases your chances for profits since the pasture, animals, fences, water, and management time are the major costs on your farm. Sustainable beef production may be an option for you because it focuses on the long-term health of the environment, while making the farm economically viable and addressing social issues that you consider important.

CONVENTIONAL FARMING

Many conventional farming practices are geared for maximum production by using large amounts of outside resources to produce their products. Whether it is corn, soybeans, alfalfa, wheat, or any number of other crops, intensive farming practices have become the norm for many farms.

These farms typically use chemically based products to control insects, pests, and weeds, and to promote rapid growth of crops and animals or to increase milk production. Hybrid seeds or GMOs are sometimes planted to increase yields. Highly specialized machinery does most of the work and the operator's feet may seldom touch the ground.

Conventional farming differs dramatically from organic and sustainable farming in that there is a heavy reliance on nonrenewable resources, such as fertilizers, gas, and diesel fuels, and the concern that some practices, such as excessive tilling, can lead to soil erosion that may cause long-term damage to the soil.

OTHER CONSIDERATIONS

One of the most attractive aspects about all three farming systems is that you can stop one system and switch to another system, although it is easier to switch to conventional farming from one that is organic or sustainable than the other way around. This may be important if a change is required in your farm somewhere in the future. You will have the option to alter your farming plan to adjust to the new circumstances.

CHAPTER 9

STOCKING GUIDELINES AND START-UP ECONOMICS FOR YOUR BEEF OPERATION

Well-grown pastures can supply the majority of feed for cattle. Grasses and legumes can be cut and stored for future use when cattle are on other pastures.

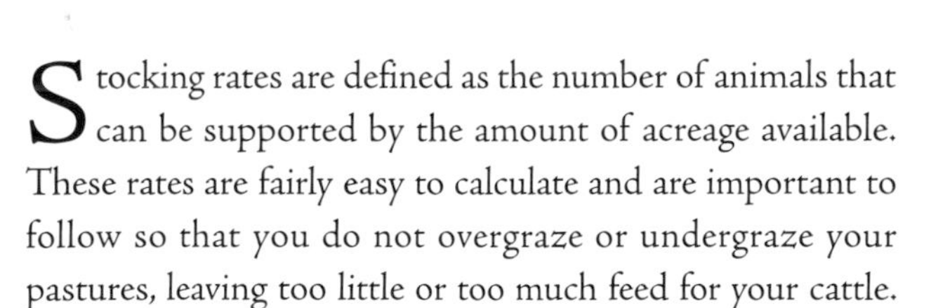

Stocking rates are defined as the number of animals that can be supported by the amount of acreage available. These rates are fairly easy to calculate and are important to follow so that you do not overgraze or undergraze your pastures, leaving too little or too much feed for your cattle.

If you plan to purchase a substantial amount of supplemental feed, such as hay and grain, to provide the necessary amount of forage for the animals' diets throughout the entire year, stocking rates may not be as important to you. However, you can keep down the costs of buying extra feed by not putting more animals in a pasture than the grass can support.

Each animal has nutritional requirements that must be met by eating grass, hay, grain, crop stubble, or whatever else is growing in the pastures or fields being grazed. Your location will determine whether you use year-round or seasonal grazing, or whether you harvest hay during the year and store it for winter feeding. Your stocking calculations must take into consideration the size of the animals because mature animals eat more hay and grain than younger, smaller animals. Although there seems to be little difference between breeds regarding their nutritional requirements, some breeds appear to have a better rate of gain from similar quantities of forages. Weather may enter into these calculations. Colder climates generally require greater amounts of feed than warmer climates.

Even after making reasonable calculations and projections, it is advisable to have a supplemental or alternative plan should you fall short on your feed supply before the new pasture crop arrives in the spring. Understanding the quantities of feed it will take to adequately supply each animal during the year will help you plan for such situations as a dry

Dividing pastures into smaller areas called paddocks allows cattle to heavily graze a certain area for a short period before moving them to another paddock. This is called rotational grazing. By decreasing the size of the paddock, cattle are more concentrated and will eat more off the tops of the plants. This keeps plants from maturing too quickly, especially during rapid growth in the spring.

year, when there is no crop to harvest or you harvest a crop that is insufficient to meet your winter needs, or a wet year, when you may not be able to harvest in a timely fashion.

PASTURE HEALTH

A successful cattle raiser becomes an expert grass grower because the quality and health of your pastures will determine the number of animals they can support. Learning how to develop and maintain quality pastures will decrease the amount of purchased feeds and increase profits. You can develop a sustainable stocking rate by having a thorough understanding of the pasture's approximate forage yield.

Typically, well-grown pastures of moderate density or thickness will produce between 2,000 to 2,500 pounds of total dry matter per acre (DM/a) in the first 8 inches of growth. This kind of growth under normal weather conditions will produce roughly 5 tons of forage per acre during the growing season.

The pasture regrowth or rejuvenation during the season will be determined when cattle are removed from a particular pasture. It is best to move them from any pasture when they have grazed the grass down to a height of 2 to 3 inches. Grazing to lower amounts greatly increases the recovery time of the plants for the next rotation.

PADDOCK SIZES AND STOCKING RATES

Paddocks are small enclosed areas located within the whole pasture area and are generally divided off with electric wire to control cattle access to parts of the pasture at one time. Dividing the pasture into small areas helps manage them more intensively because your cattle are allowed to eat in a

Cattle paths are the result of heavy foot traffic along fences and through other parts of the pasture. Seeding these areas in late fall or early spring helps the regrowth of grasses.

specified area for a short time and trim down the growth before they are moved to another paddock to graze.

Unless your cattle are free-range (meaning they have access to the entire farm acreage that you use for your pasture grazing program), you will need to develop paddocks to help manage your whole system. The size and number of paddocks will be determined by the number of acres you commit to your grazing program, the health and density of the pastures, and how fast the pastures regrow. Unfortunately there is no set answer for your system. The best way to determine when to move your cattle is to walk out into the pasture each day and look at what's there and how much they've grazed.

This is where experience and learning will set you apart from conventional farmers. You will learn to read the land and the grass in ways you may never have thought possible. Even experienced grazers continue to work and learn with their programs to get the optimum benefits from their pastures.

Grazing capacity is expressed in animal unit months (AUM), which is the amount of forage required by one animal for one month. One animal unit for this discussion will be defined as an 1,100-pound mature beef cow that requires 30 pounds of forage dry matter each day.

This AUM is calculated for a beef cow with a calf by her side that requires 4 percent of her body weight through the year. If you are pasturing dairy cows, roughly the same poundage per day is required.

CALCULATING PADDOCK SIZE

You can use the following formula to calculate the size of a paddock needed and account for the daily dry matter requirement. This formula (ar right) is roughly the same for a beef cow or a lactating dairy cow:

Rapid movement of cattle through a paddock system allows them to consume the most nutritious portions of the plants and better utilizes the pastures. By moving cattle frequently, you will be able to stay ahead of the rapid plant growth. When plant growth slows during the summer, paddocks can be enlarged and the cattle can be moved less frequently.

Round bales may be stored in sheds or stacked out in the open and covered with tarpaulins to prevent spoilage. Round bales will shed rain and snow because they are tightly packed. However, the best feed quality is preserved if they are stored under cover.

(No. of animals) x (DM intake) x (Days in paddock)
÷ lb forage/acre = paddock size

As an example, if you have 25 beef cows x 30 lb DM
x 2 days ÷ 1,200 lb/acre = 1.25 acre paddocks

The figure of two days used for the amount of time the animals are allowed in a specific paddock does not take into consideration the very early season when the grass is starting to grow or late spring when the pastures are rapidly growing. During these times it is advisable to move your animals through each of the paddocks quickly to even out their consumption. Moving cattle quickly through the entire system in times of rapid plant growth will allow for the removal of the tops of the grasses, which deters them from forming seed heads and maturing too early and lowering the pasture quality.

If a calf is alongside a cow, these figures will be somewhat higher but not significant enough to make a real difference. As a rule of thumb, one cow-calf pair should be allocated 1.5 acres for the year. However, this figure will vary depending on which part of the country the land is located.

CALCULATING STOCKING RATES

To use another example, assume you have 50 acres of pasture available to graze. Also assume that you want to place as many beef cows on your farm as possible and that you want to harvest some of those acres for dry hay to be stored as bales to be fed during the winter. You also don't want to purchase any outside hay for your cows. The number of cows this acreage will support can be calculated as follows:

(1 cow x 30 lb/day x 30 days/month
= 900 lb DM/month x 12 months
= 10,800 lb DM/year

then,

5 ton DM/acre over growing season
= 10,000 lb DM x 50 acres
= 500,000 lb DM/year produced from the pastures

then,

500,000 lb DM/year ÷
10,800 lb DM/year/cow = 46.3 head

These calculations show that this 50-acre pasture could potentially support 46 beef cows during a normal year. Part of this acreage could be harvested for hay to be stored for winter feeding. These numbers do not reflect the

Transporting cattle is easy with a low-clearance trailer. Animals can be loaded quickly and safely because of the short height they need to step up into the trailer. Many custom-hire haulers use this type of vehicle and trailer.

calves that will grow during the year and will start to consume their own grass and dry hay. The total number may be several head less than projected.

In perfect conditions these numbers look very good. However, a 46-head beef herd is a significant number of animals to care for. Considering the numbers used for your paddocks, it is apparent that the size of each paddock would need to be doubled to accommodate twice as many head as used in the example.

These are rudimentary examples that may or may not fit your situation but give you some idea how to calculate the number of head your pastures will support. Also, the body size of the animals may require a slight adjustment in these numbers as well.

These numbers can be adjusted if you decide to purchase hay from another source or if you use silages, such as corn, oats, wheat, or any other grasses that can be stored and cured. Weather conditions during the year will certainly play a role in these calculations that may need to be readjusted as the season progresses.

CALCULATING STOCKING COSTS

Every farming enterprise has beginning costs, and understanding these costs will help eliminate surprises in setting up your beef project. Some cattle-stocking costs may rise or fall depending on market conditions at the time you

Metal feed troughs placed along fence lines provide a space to put mineral and salt blocks for cattle on pasture. Free choice minerals and salt should be available at all times.

buy your animals. The price of the animals you purchase may depend on such factors as the time of year, feed availability in a given area, or the breed of animal you choose.

Having calculated the number of animals your farm can support, you can reasonably calculate the total cost of buying your animals. While market conditions may change from the time of this writing to when you purchase your animals, here are some figures to guide your estimate of how much investment is required.

1 cow = $750—single cows may be cheaper but be sure they are pregnant
1 cow & calf = $850—a cow with a calf can be a worthwhile investment as you get two animals for the purchase price
1 calf = $100 if you are starting out as a feeder calf farm

These numbers can then be multiplied by the number of animals you intend to purchase.

Metal feeders are used for feeding round bales, small square bales, silage, or loose hay. Place them in farm yards, corrals, or pastures with easy access.

There are other considerations to make when buying animals, such as: Where do I find animals? Are transportation costs involved? What health issues do I consider? What do I use for animal identification?

BUYING ANIMALS—PUBLICLY OR PRIVATELY

There are two avenues through which you can acquire animals for your enterprise: public auction or private purchase. Each has its advantages and disadvantages.

A public auction is where any member of the public can bid on animals to purchase. In these auctions you will pay a price that is the highest bid and you will be expected to present cash or a good check after the sale. One advantage of buying at a public auction is that there are established conditions under which the animals are sold. Sometimes guarantees are stated at the beginning of the sale regarding the animals. These guarantees could include pregnancy information, vaccination status, and a variety of other details that may be important to you.

Buying cattle at public auction is one way to acquire animals. Your good judgment of an animal may be confirmed by the number of other bidders.

Large square bales can weigh 1,000 pounds and have a similar volume as large round bales. Square bales are easier to store but proper equipment is needed to handle them.

Another advantage of auctions is that the price is determined by other bidders. This can provide a reasonable assessment of the worth of those animals by other, perhaps more experienced, farmers and can affirm your judgment of the animals. In some cases the price may be more than you want to pay and you go home empty-handed. At a public auction the animals generally sell one at a time unless a group of animals is being sold together. This may mean that you will have to attend several sales to get a number of animals.

The final bid is the price you pay and there are usually no negotiations at the end of the bidding. Unless there is a major problem discovered after the purchase, you generally can't return the animals. However, that doesn't mean you can't discuss the problem with the seller at a later time to arrive at an agreeable solution. Most established breeders are willing to make things right if problems arise, especially if they know you are a new customer. It is in their long-term interest to have satisfied customers.

With a private purchase of animals, you may pay a price that is determined by the owner, but you also may be able to get all your animals at one place without having to spend a lot of time going to different auctions to acquire the number of animals you need. When buying privately it may be possible to negotiate a better price if a large number of animals are involved in the purchase. An established breeder's reputation is important and most will try to accommodate a buyer's concern. You may also be able to get replacements for those animals that develop problems.

TRANSPORTATION COSTS

One overlooked cost of buying animals privately or at public auction is the issue of transporting the animals to your farm. If you do not have a cattle trailer to use, alternative transportation, such as hiring a trucker, will need to be arranged.

As the buyer, you will be expected to promptly remove the animals from a public auction. If you can hire a trucker to haul them to your farm, make certain you understand the driver's terms for hauling and get a firm estimate. Usually the cost is based on each mile of transport. Depending on the individual truck driver, you may be quoted per head or by the load. If hiring transportation for your purchases, insist that the trailer is thoroughly sanitized before loading your animals.

If buying privately, it may be possible to have the seller transport them to your farm. This may be part of the negotiated total purchase price.

HEALTH CONCERNS

Before the animals arrive on your farm it is important to know their health status. At a public auction some animals are sold with health certificates issued by the seller's veterinarian. This certificate is important because it assures that the animal is healthy and as reasonably free of disease as is possible to determine.

If the animals are purchased privately it may be possible to secure health certificates from the owner prior to leaving the farm. Be sure to fully understand who pays the cost of having the health tests done while the animals are still at the seller's farm.

The simplest way to quickly determine the health of an animal is to look at it. Does the animal appear to be alert? Does it have clear, dry eyes? Thin or emaciated animals should be avoided entirely, no matter what the price, because this may indicate serious health problems. Avoid animals with runny noses, discharges from their eyes, breathing abnormalities, or anything else that strikes you

Large square bales feed many animals at one time when placed in a metal feeder such as this. With head locks, this type of feeder is versatile for grazing programs and allows free access to dry hay.

Hay can be harvested while it still contains moisture. It is wrapped tightly with plastic to create a silage-like material. The moist hay ferments inside because the plastic holds in the moisture and keeps out air. Bale units can be opened individually as needed to reduce spoilage.

as not being normal. If you are not sure of your own expertise at identifying problems, hire a veterinarian or someone with experience to go along to look at the animals prior to purchase; this will be money well spent.

ANIMAL IDENTIFICATION

Identification of the animals you have purchased is important so you can be assured the animals that arrive on your farm are the ones you purchased. If the animals have ear tags of any kind, take note of what is printed on them because this can verify them as the ones you purchased.

Some form of identification, whether it is by metal, plastic, or radio frequency ID ear tags; tattoos; photographs; freeze brands; retinal scans; neck chains; or some other semipermanent or permanent method will help you keep track of the animals you work with through the years. Consider using ear tags or some other form of identification that can be attached to the animal before or shortly after they are unloaded at your farm. Record-keeping will be easier if you can positively identify each of your animals.

CHAPTER 10

FEEDING—BEEF CATTLE NUTRITIONAL REQUIREMENTS

Stocking guidelines will help you determine the number of animals that your farm can support. Understanding the nutritional requirements of your cattle will help you determine the types of crops needed to supply that nutrition. The daily nutritional requirements of each animal during the growing season can be met using grasses in a grazing program. Raising cattle on pasture-based diets is gaining popularity with beginning farmers because of lower feed costs associated with a grazing program and because the farmers are dissatisfied with conventional programs.

Silos are used to store forages for use when pastures are short, or during the winter when pastures or forages are no longer available. Grasses, legumes, and corn can be stored for long periods of time in silos because the material ferments as it settles and seals in the nutrients. Silage is removed by using mechanical unloaders.

RUMINANTS

Cattle are ruminants, which means their stomachs have four chambers. The chambers are basically fermentation vats that break down plants and grain into carbohydrates, proteins, enzymes, and other components needed for growth, maintenance, or milk production.

This specialization gives cattle the ability to absorb nutrients and break down grasses through acidification. This allows ruminants to extract nutrition from low-quality feeds and to efficiently utilize plant products that other animals cannot use. In this way they become a conduit for products not edible by humans and convert them into food, such as meat and milk, that humans can use.

The key to the ruminant digestive system is a symbiotic relationship between the microorganisms that exist in the fore part of the stomach. These bacteria and protozoa have the capability to convert solid plant material into enzymes usable by an animal. As these microorganisms work together, they produce enzymes called cellulase, an enzyme that no animal can produce on its own. Cellulase enzymes break down the cell walls of plant materials,

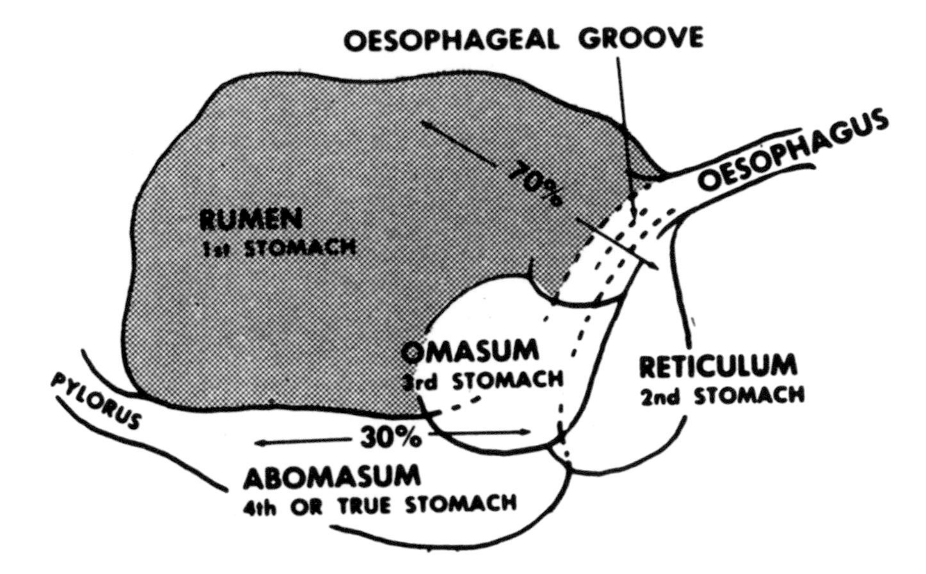

Diagram of a cow's digestive system showing the four stomachs.
Nestle Carnation-Albers

which releases the plant's fatty acids into the digestive system. These fatty acids are then absorbed by the cow and make a significant contribution to her overall energy needs. As the other plant materials pass on into the intestine, amino acids, lipids, carbohydrates, and other enzymes are absorbed in much the same way as other animals.

The use of microorganisms to aid in two different digestive systems allows a cow to exploit cellulose as an energy source that other plant-eating animals cannot. This makes them ideally suited to eat crops produced on poor soils that may be indigestible or unsuitable for other species.

THE DIGESTIVE SYSTEM

The four chambers of the cow's stomach include the rumen, reticulum, omasum, and abomasum that work together as a specialized system. The rumen is the largest chamber where the grass eaten by the cow first enters the digestive system. The rumen becomes filled with chewed and half-chewed materials that are mixed with saliva as the cow regurgitates and swallows the materials several times. The chewing process grinds the food into smaller portions and injects saliva into the material as it is crushed and chewed. This crushing allows more surface area of the material to become exposed to the bacteria in the rumen.

As the fiber breaks down into cellulose, the bacteria and protozoa break down the cellulose into cellulace and then into glucose, which is used by the microorganisms to feed themselves. While some microorganisms escape the rumen and pass through the other chambers, most stay behind to work on the newly ingested plant materials.

The fiber that is broken down immerses itself into the liquid portion (saliva and stomach acid) and then passes through the rumen to the next chamber, the reticulum, and then to the omasum, where the water is removed. All through this process, the materials are constantly being churned to mix the liquids, solids, and bacteria to keep the fermentation process going.

As the water is being removed, the material moves along into the next chamber, the abomasum, where it becomes digested much like it would be in the human stomach. The abomasum, unlike the rumen, reticulum, and omasum, does not absorb nutrients. It prepares food for enzymatic breakdown and absorption in the small intestine.

From the intestine, where the absorption of nutrients continues, the unusable portion is finally expelled as manure. The digestive process of the ruminant produces acetic acids, propionic acids, and butyric acids, which are the volatile fatty acids produced by the bacteria that give the ruminants their energy.

GRASS FOR YOUR CATTLE

Cattle can obtain most of the nutrients and energy they need for growth and production from good-quality grasses alone. Grasses, clovers, and alfalfa are the most abundant natural resources available for feeding cattle and are the least expensive crops to produce and harvest. In many areas of the country, grass grows plentifully and can be used as a major source for feed and roughage.

Grasses can grow virtually everywhere, particularly in the Midwest. Ironically, grasses are so prevalent and hardy that many farmers spend huge sums of money trying to eradicate grasses from their fields with herbicides and other chemicals to control their growth.

Grasses can be used to your advantage and there are a number of varieties available to provide the nutrients necessary for an adequate rate of gain by your animals. Understanding the basics of plant growth is essential to establishing and maintaining pastures that adequately support your animals for the full growing season.

PLANT GROWTH—THE KEY TO GOOD PASTURES

Photosynthesis is the process that transforms sunlight into the energy plants need for growth. This energy is converted to carbohydrates, which can be used for growth or stored for future use. Because photosynthesis occurs in the leaves of plants, the plants grow slowly at first due to the small surface area of the leaves. This is evident in the early spring or after a cutting of grass has been removed or grazed off. This is called recovery time and during this phase the plant is using some of the stored energy in its roots to start growing again. As the leaves enlarge, photosynthesis increases dramatically and the plants grow rapidly. As the plant matures, its growth slows as it develops flowers to produce seeds. At this point, just before the flowering stage, the nutrient quality of the plant reaches its peak.

The quality of the plant material decreases as the plant matures further and diminishes its nutrient content. As the plant ages, a greater percentage of its nutrients become tied up in nondigestible forms, such as lignin. As the amount of nondigestible fiber in the plant increases, the result is lower quality forage with a decrease in total digestible nutrients (TDN). TDN is used in feed ration calculations to help determine the contribution of the forage to the animal's dietary requirements.

One way to circumvent the problem of fast maturing, low-quality forage is to quickly move the cattle through the pastures, perhaps as often as twice a day. This will allow cattle to eat many of the tops of the grasses that will push them back until they can recover through regrowth.

Since cattle need feed year-round, it is difficult to provide an adequate supply from pasture alone, particularly in regions with an extended winter climate such as the

Silage bags are used on many farms and are sealed at both ends when filled. The forages ferment the way they would in an upright silo. One advantage of silage bags is the ease of loading large quantities of silage in a short period of time and there are no mechanical unloaders. The disposal of large quantities of used plastic may be difficult.

Silage mounds can be used if no other storage facilities are available. The material is packed down by a tractor as it levels and shapes the pile. It then can be covered with a tarpaulin to seal it for fermenting. One disadvantage may be moisture that seeps in from the bottom of the mound if the forages stored are too wet when harvested. Any mound should be located away from waterways or streams so runoff is not a problem.

Northeast, Upper Midwest, and many Western states. By leaving several fields for harvesting or following behind the cattle and cutting the grasses they leave, you can store forage for winter feed. This can be done as dry baled grass and hay, or as silage when put into a storage facility, such as a silo, bunker, or bag. Harvesting grass will allow these areas to regrow at different rates from other paddocks and may assist in your rotation plans. It will also give you a hay supply for winter feeding.

TYPES OF PLANTS

Many different varieties of grasses and legumes are available to supply the nutrients needed by your cattle. A pasture with a mixture of several grasses is preferable to one that contains a stand of only one variety. Having multiple varieties growing in a pasture allows more flexibility in climate conditions occurring during the year. Each species has different strengths, so planting several varieties in a pasture will provide more stability in changing conditions.

Bunker silos are popular on large farms because of the ease in unloading trucks that haul forages to the site. The piles are tightly packed as the tractor levels the forages as it drives over the feed several times.

Tractors with huge loading buckets are often used to scoop silage and put it into mixing tanks. Cement bases in bunker silos provide a clean, solid area for loaders to operate.

Some grasses, such as timothy, clover, and bluegrass, prefer cooler temperatures and are more productive in the spring and fall. Legumes, such as alfalfa, start growing a little later in the spring but have a uniform growing pattern during the summer season. There are grasses that thrive on the midsummer heat, which normally slows down many other types. These are warm-season grasses and include bluestem and switchgrass, which tolerate very dry conditions.

Some grasses have very good quality in the spring and can be used for pasturing if their seed is not needed later, but they are not very palatable as they mature. These include oats, wheat, and winter rye, except if it is sown in late fall and the growth is enough that it can be grazed before winter. These grasses provide excellent feed in their early stages, but as they mature their quality drops dramatically after the seed heads mature.

The species of plants used in your pastures should be tailored to the grazing system you have designed for the region where you live. The climate, soil type of your farm, amount of moisture your area typically receives each year, length of your growing season, and the number of animals you want to put on your pastures are all considerations for the kind of grass and legume species seeded into your pastures.

There may be no single forage best suited for all farms but some of those varieties that can be incorporated include:

Alfalfa is the highest-yielding legume in many states. It has excellent summer regrowth and can do well in some dry weather conditions. It does not tolerate overgrazing well and if it is a pure stand in the field, bloat in cattle can be a potential problem.

Birdsfoot Trefoil can be difficult to get started, however once established, it grows well on poor soils. Because it can maintain its quality better than other legumes, it is well-suited for stockpiling during late fall.

Kentucky Bluegrass can provide good yields but needs to be adequately fertilized. It does not do well if overgrazed or if planted on low-fertility soil.

Ladino and Alsike are white clover varieties that tend to be more persistent than red clover. Both do well in pastures that get heavily grazed and although they have lower yields, they have a higher quality. Alsike does well in wet areas of pastures or fields.

Orchardgrass has a high yield and because it is extremely competitive, it recovers quickly and will challenge legumes for space.

Perennial Ryegrass is easy to establish, which makes it a good choice for a quick temporary pasture. It has high yields and quality. One disadvantage is that it tends to die out after a couple of years.

Quackgrass can be an excellent feed source for cattle because of its high yield and quality. However, it is the scourge to those who consider it a weed in row crops.

Red Clover is good for a temporary pasture. It has high yields, grows fast, and is easy to establish. Bloat can be a concern if too much is eaten too quickly.

Reed Canarygrass can be a mixed blessing. It is very useful in wet, poorly drained areas or places where it frequently floods. Although it is difficult to establish, it is extremely persistent and can displace native plant populations. Some varieties can be low in palatability.

Smooth Bromegrass is easy to establish and has good regrowth potential if grazed or harvested early. As it matures it tends to be better for harvesting as dry hay for the winter.

Switchgrass and **Big Bluestem** are warm-season grasses that complement cool-season grasses. Although they

Mixing wagons are used to deposit silages and grains from tractor loaders. These wagons stir the contents in much the same way as a cow's stomach mixes the different grasses and grains she eats. By mixing these feeds for the cow, she needs less time to mix them herself and the result is more milk as well as a better rate gain in beef.

both are slow starting, they are persistent once established and will be vigorous growing plants. Big Bluestem is higher in quality but generally is higher in seed cost.

As a rule, remember that it is better to have a mix of different grasses in each pasture rather than a single stand of one variety. Typically a mix of alfalfa-brome-timothy or clover-timothy or some other combination of grasses that grow well on your farm will provide sufficient forage during an average growing season. Hedging your bets with a combination may prove to be worthwhile during years of abnormal weather conditions.

FORAGE NEEDS OF YOUR CATTLE

Having a clear understanding of the total forage needs of your cattle will help you establish a grazing program that uses your pastures to their best advantage. This will help minimize the amounts of unused forages that are left in the pastures or shortages during the grazing season.

CALCULATING FORAGE NEEDS

The University of Wisconsin Extension has developed a formula for approximating the forage needs of your cattle. This formula can be used each month to calculate an estimate of production throughout the season. However, this formula does not take into consideration differences in animal needs during the season or the pasture production at different times of the year.

Season average animal weight x daily allowance x number of days on pasture = total pounds of forage needed

For example, if you have 20 beef cows weighing on average 1,000 pounds each and you want to put them on pasture for six months (180 days), April through October, and assuming the daily allowance requirement is 4 percent of body weight (as identified in chapter 9), then:

20,000 x 0.04 x 180 = 144,000 pounds or 72 tons

Typically a good pasture will produce from 3 to 5 tons of forage per acre per year. You can determine the amount of acreage needed to supply the 72 tons for the grazing season. Mechanical harvesting will provide a better and more total cut on the field so that grazing will not give the optimum total supply of feed. Some forage will be left behind as the cows move from field to field or trample down some of the plants while grazing.

If you have 50 acres to devote to your grazing system that produce 4 tons per acre per year then,

50 x 4 = 200 tons forage per year

If your pastures produce at this level there should be sufficient acreage for your cattle to graze during the summer and still harvest enough to feed them through the winter.

OTHER CONSIDERATIONS

During times of stress on the pastures, perhaps due to dry conditions, some grazers will use grain as a supplement to the grass available. Most of this grain mix consists of corn that is mixed with minerals and perhaps some vitamins, depending on the condition of the soil fertility and the plants. Using grain as a supplement to the grass can have some beneficial effects for rate of gain. However, it may not fit your goal of a grass-based grazing program.

If pastures are insufficient for raising the number of animals you have, one option may be to cull the herd by selling the older cows first and then selling the young stock as feeders. If purchasing hay, silage, or other forage is within your budget, it may be an option to buy this feed and continue to raise the young stock. However, if you wish to feed grain as a supplement to hay or pasture there are ways of determining how much benefit you can gain from it. Typically, most grass hay has only 50 to 60 percent the energy content of grain, so 1 pound of grain can replace 1.5 to 2 pounds of hay.

In a case such as this it would make very little sense to pay $100 per ton for poor-quality grass hay when the same amount of grain would cost just a little more. If you decide to incorporate grain into the diet of pastured cows, it is important to start them on grain very slowly and be sure all cows have an equal access to the grain. Pushing a cow's diet with grain too quickly can lead to acidosis and other intestinal problems because the stomach flora has not had the chance to adapt to the grain that requires different bacteria to break it down.

STOCKPILING

The purpose of stockpiling grass is to provide feed for cattle during the winter and in early spring (before the plants begin to grow). Leaving some of the grass at a good height prior to winter will enable you to offer your cattle an early pasture feed source. Stockpiling requires cattle to be off these pastures prior to grazing them below 5 to 6 inches in height. Some shrinkage will occur during the winter but there usually will be enough for them to graze when turned out early the next year.

Bunker silos are constructed in sections to allow for the storage of different feedstuffs in each section. Feeding from one section until it is empty will reduce loss of quality and prevent excessive spoilage.

Small square bales weighing 45 to 60 pounds may be baled instead of large square or round bales. The smaller bales are easier to handle and may fit better in your storage areas.

FORAGES REQUIRED FOR RATE OF GAIN

Achieving a desired rate of gain requires meeting the nutritional needs of the young animals. Your pastures can supply these needs as well. For example, assume that your goal is to raise a calf to 1,100 pounds for market. Assuming the initial nutrition for the growth of this calf will be furnished from its mother's milk through the first five to six months until the calf weighs 500 pounds, this leaves 12 to 13 months during which the forages must provide the rest of the growth.

A 2-pound-per-day rate of gain on pasture is quite feasible and sometimes it can increase to 2.5 pounds per day. A rate of gain of 2 pounds gain per day will, over the course of 12 months, raise a calf to an 1,100-pound animal.

How much forage is needed to reach a 2-pound-per-day rate gain? As a rule, it takes about 10 pounds of forage to produce 1 pound of gain in beef. If the majority of the growth after six months comes from grass pasture then,

1,100 – 500 = 600 pounds of gain needed to reach market weight

600 ÷ 365 days/year = 1.64 pounds rate of gain

This is well within the parameters you have set. If the calf requires 4 percent of its body weight to be eaten each day, then a 500-pound calf will consume 20 pounds of dry matter each day.

365 days x 15 pounds/day = 5,475 pounds of dry matter, or 2.5 tons, or about 1/2 acre pasture per year

Grasses and legumes harvested during the growing season are an excellent source of winter feed for cattle. The wastage pulled onto the yard can be used for manure in spring.

CHAPTER 11

REPRODUCTION—REPLENISHING YOUR BEEF HERD

A "heat" posture is exhibited by one animal riding another. The animal in heat stands to be mounted. This is the most obvious sign of heat. Other actions by animals prior to and after exhibiting this posture may include head butting or attempts by the cow in heat to mount other cows. When no bull is running with the cows, artificial insemination should take place within 6 to 8 hours of this sign to realize optimum pregnancy rates.

Replenishing animals in your herd through a successful reproduction program will enable you to keep the number of animals needed for your farm at a sustainable level. Understanding natural reproductive processes will enable you to utilize those animals you have chosen, regardless of breed, to repopulate your herd even after selling some of the animals over the course of the year.

TERMINOLOGY

There are basic terms that refer to physical differences in the reproductive capacity of animals that need to be understood, especially if your prior experience with cattle is limited. These are terms normally encountered in the day-to-day conversation with other beef or dairy producers and apply to any of the bovine species.

Cow: Female

Bull: Male

Steer: Castrated male

Heifer: Female under one year of age

Yearling: Female between one and two years of age

Heat: Estrus; the period of sexual excitement in cattle and the time for breeding

Artificial insemination (AI): Process of using frozen semen to impregnate a female

Breeding: The act of depositing semen into the female reproductive tract

Freshen: To have a calf and come into milk production

Settle: Refers to a cow having conceived and not returning to heat

These are not the only terms you will encounter but they are those most pertinent as you begin to establish your herd.

REPRODUCTIVE ANATOMIES

Cattle have a similar reproductive structure to other mammals. The position and utility of the reproductive organs serve the same functions. The following diagrams show how the reproductive organs of a cow are situated. Understanding the positioning of these organs will help when breeding your cows by artificial insemination or when using a bull for natural service.

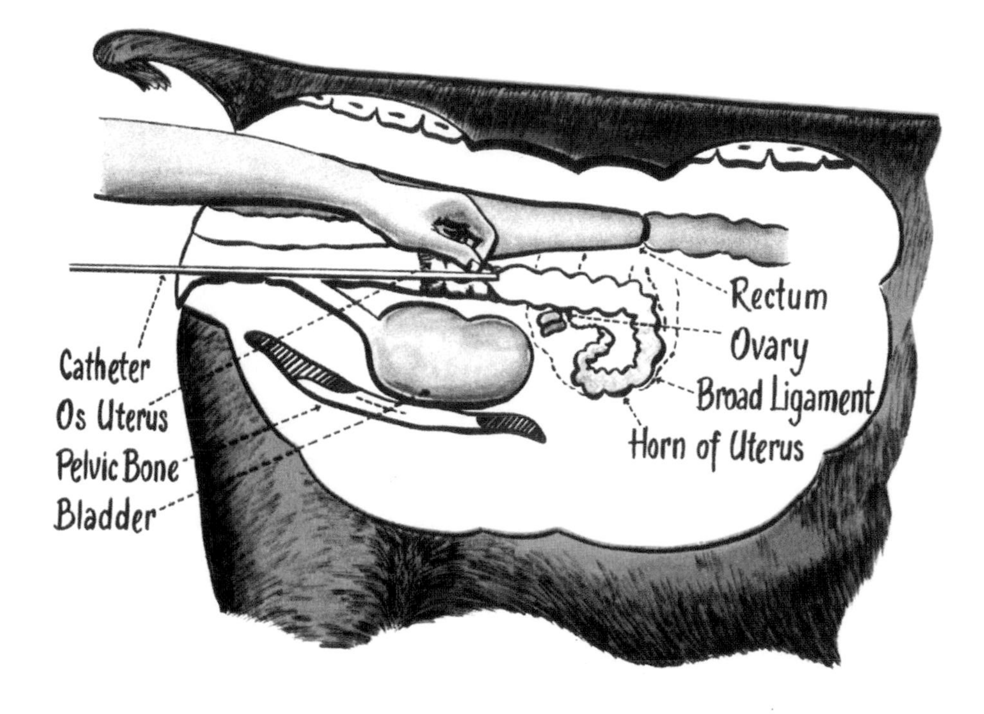

This diagram shows how cows are artificially inseminated. *Wisconsin Agriculturalist*

Normal position. *Wisconsin Agriculturalist*

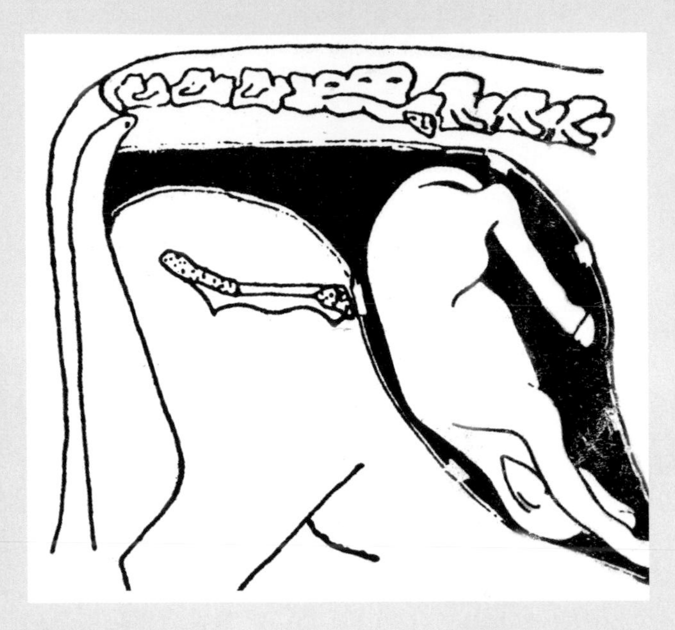

Backward and upside down. *Wisconsin Agriculturalist*

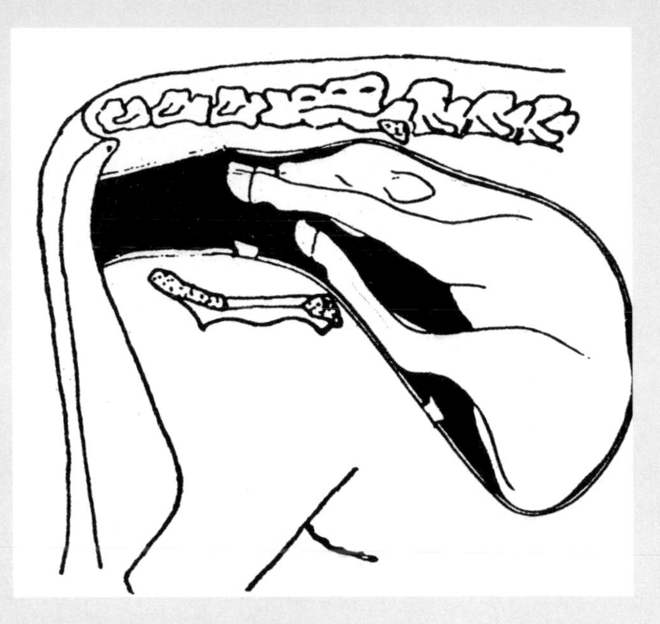

Head first position with rear legs under the body. *Wisconsin Agriculturalist*

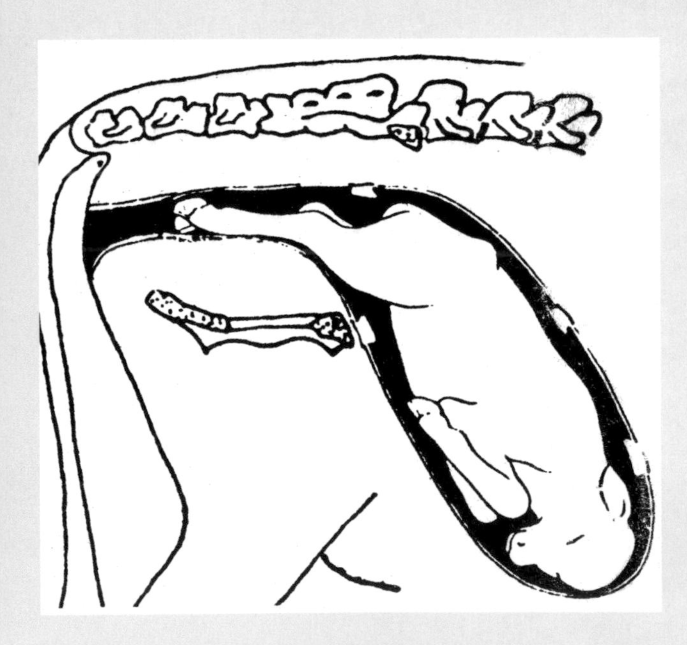

Hind feet first—right side up. *Wisconsin Agriculturalist*

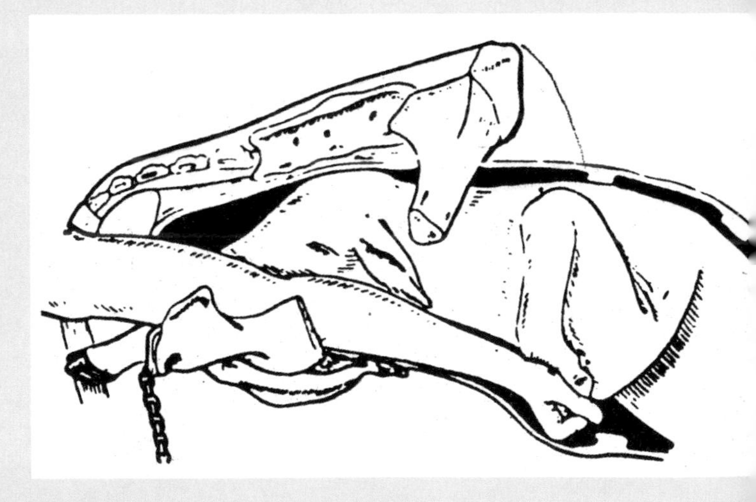

Simple leg flexion in head first position. *Wisconsin Agriculturalist*

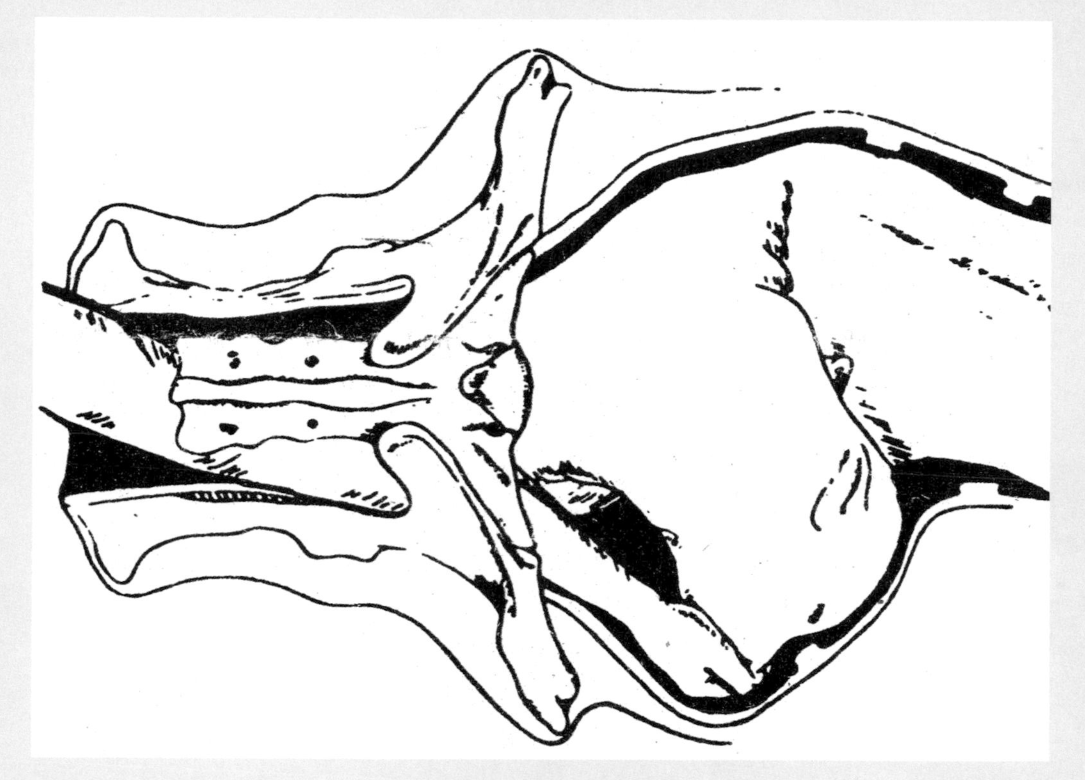

Head first position with head and neck turned back.
Wisconsin Agriculturalist

Hock flex—hind feet first.
Wisconsin Agriculturalist

ESTRUS

The female reproductive system has two ovaries that lie within the pelvic cavity of the cow or heifer. The ovaries are similar to a bean in size and shape. Both are intimately related, working in conjunction, and are attached to the tissues that envelop most of the other reproductive organs. When the cow is open, that is not pregnant, the ovaries lie within or on the front edge of the pelvic cavity. In advanced pregnancy, the ovaries are carried forward and downward into the abdominal cavity along with the enlarged uterus containing the fetus.

Ovaries, through the production of ripe ova, or eggs, set the pace for the rest of the reproductive tract that must be adjusted to receive the fertilized egg and carry the developing fetus through to birth. The ovaries have a dual function: the production of eggs and the secretion of hormones. The hormones cause necessary adjustments in other parts of the tract to take place. The heat cycle or rhythm is divided into several well-marked phases.

The first phase is the development period, or proestrum. During this phase the Graafian follicle within the ovary becomes larger, principally due to an increase in fluid. At this point, the follicle resembles a water blister on your finger. The egg develops within the follicle, along with a hormone in the fluid called estradiol, which is estrogenic, or heat-producing. The estrus or heat stage follows as the

A frozen semen storage tank has a vacuum-sealed outer lining that keeps the liquid nitrogen stable. Tanks should be refilled every 6 to 8 weeks to prevent them from going dry as liquid nitrogen evaporates over time. Companies that provide bull semen offer this service.

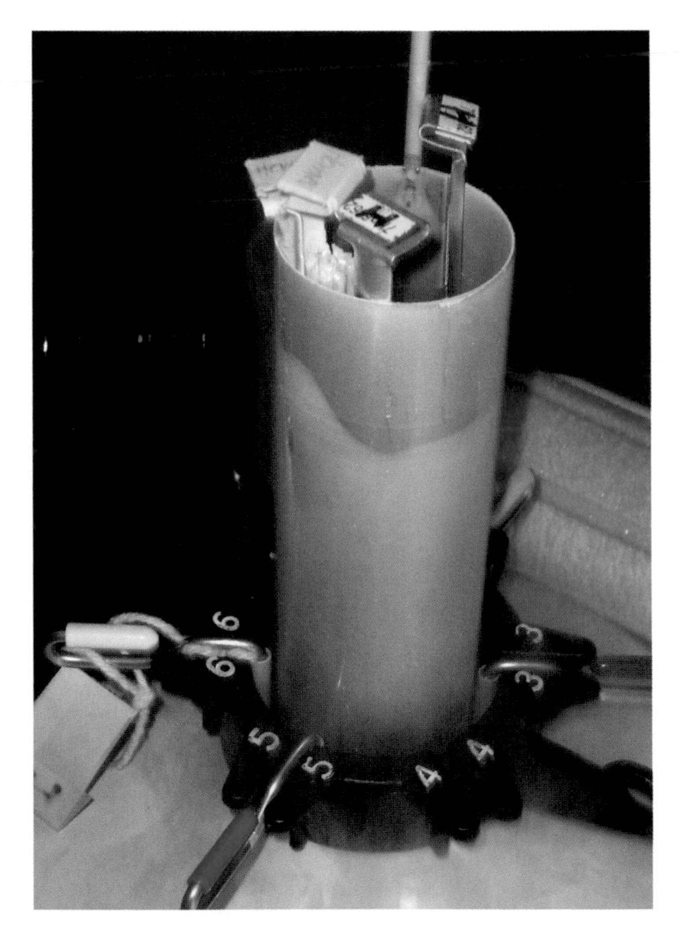

Six canisters are typically found in a semen tank and can hold many units of frozen bull semen. These canisters sit deep in the tank and are immersed in liquid nitrogen to maintain their quality. Labeled canisters provide easy management of the semen inventory.

estradiol enters the blood stream of the female. This is the period of desire and acceptance of the male and is well-marked in cattle by the standing posture.

The follicle in the ovary ruptures after the heat period has passed. The average time of ovulation is about 10 to 12 hours after the end of heat. If using artificial insemination, this is an important factor to consider. If the cow is bred too early in the heat period, her chances of conceiving are less than if she is bred late in the heat period.

After the follicle has ruptured, the egg travels down the fallopian tube to the uterine horn where it is fertilized with sperm. The former follicle cavity transforms into the corpus luteum, which produces progesterone that prevents further heat periods from occurring and interfering with the present process. After fertilization occurs, the egg implants itself into the lining of the uterus and the fetus develops.

If fertilization fails to occur, the corpus luteum regresses in size and loses its influence on the process. This allows the Graafian follicles to develop once more and the whole cycle starts over again. Approximately 21 days later the cow or heifer will again exhibit signs of heat.

These are the basics of the reproductive process. Reading materials for a deeper understanding of this process are readily available.

DETECTING COWS IN HEAT

The most obvious signs of a cow or heifer in heat are the standing posture of the animal and a clear mucous discharge from the vulva. At the optimum time of heat, the female will readily stand in a stationary position waiting for another animal to mount her, often referred to as riding. This stance tells the bull that the cow is ready for him to perform his reproductive function by inserting his penis into her female reproductive opening, or vulva, and depositing his sperm to impregnate her.

If a bull is available, nature will complete this function readily enough. If you have decided to use artificial insemination, observing this posture in a cow will alert you to the fact that it is time to inseminate the cow or heifer. The female can be restrained so that she can be bred artificially by you or by a bull stud technician.

The semen is inserted using a hollow stainless steel tube with solid rod plunger called a gun. The gun is lightweight but durable with an O-ring at its base to hold the plastic sheath.

Each unit of frozen bull semen is identified with the registered name and number of the sire, breed, date of collection, and stud of origin.

The tip of the straw is cut off, which opens the end to allow the discharge of semen when used. Sufficient space remains inside the straw so the cutting of the end does not touch the semen.

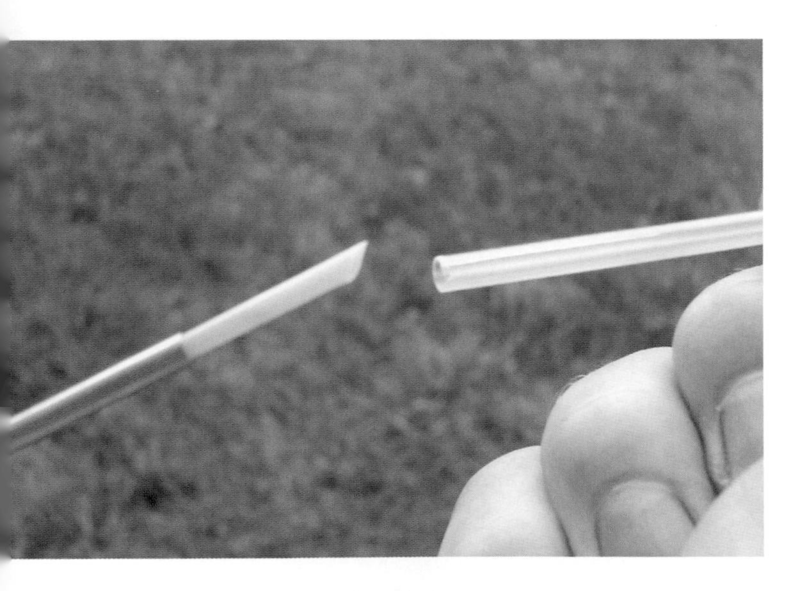

A plastic sheath slips over the straw and gun and is held in place with an O-ring at the base. This hollow sheath holds the straw and allows the thawed bull semen to pass through the end into the cow.

Most of the artificial-insemination bull studs have technicians available to breed your cows for a fee or they can deliver frozen semen to your farm prior to the time you need it. Planning and working with them will help you have the semen and supplies on hand when needed.

GESTATION

The normal range of gestation in cattle is between 260 to 296 days, although the majority will freshen between 277 to 283 days. As the gestation period advances toward the time of calving, the udder of the female will start to develop or enlarge. This is nature's way of anticipating the arrival of the newborn calf that will need milk for nourishment.

Once the calf is born, the first milk from the cow is called colostrum, which contains immunoglobulins. These are essential for disease resistance and the ability of the calf to absorb these antibodies during the early hours of its life will help protect it. The calf steadily loses its ability to absorb these antibodies over a period of hours and after about 24 hours, the absorptive ability practically disappears. Consequently, the earlier the calf receives colostrum, the greater the benefit. This is one reason some cattle growers pay particular attention to the mothering ability of the breed they choose.

ARTIFICIAL INSEMINATION OR NATURAL SERVICE

You can choose between using artificial insemination or using a bull for natural service to impregnate your cows and heifers. There are pros and cons for both processes and some of them are explained here so you will be better informed to make the appropriate choice for your operation.

Artificial insemination (AI) involves securing semen packaged in small plastic tubes called straws. Bull semen is collected and processed under sanitary conditions at a bull stud, an organization that owns bulls from most every dairy breed and numerous beef breeds. There are many of these bull studs located across the country and throughout the world.

Each 0.5 cubic centimeter (cc) straw is filled with approximately 10 million sperm, quickly frozen to a temperature of minus 320 degrees Fahrenheit, and suspended in liquid nitrogen where the semen can survive indefinitely.

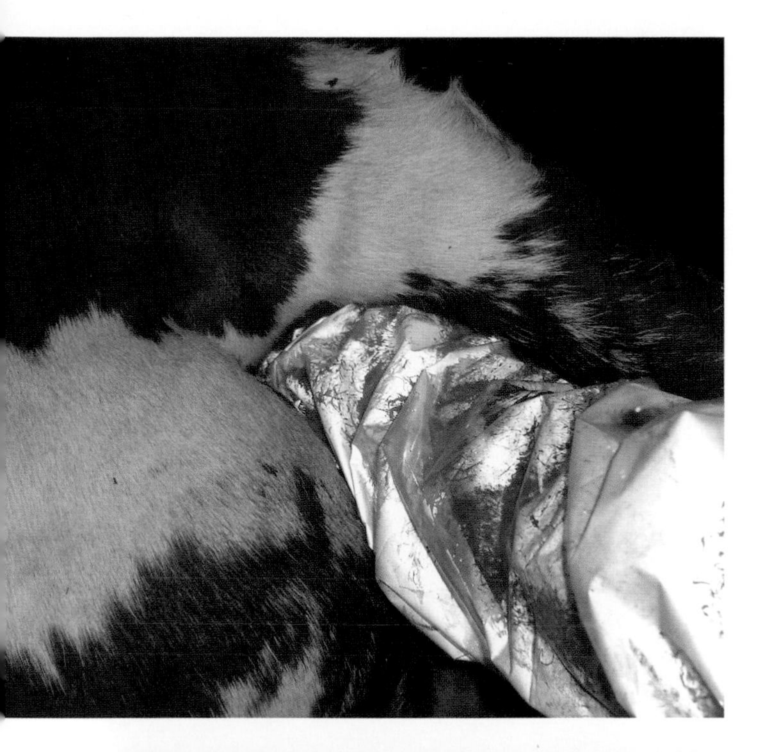

Straws come in sticks of 5 or 10 units and can be stored in a portable semen storage tank that, depending on its size, needs to be refilled with liquid nitrogen every six to eight weeks. Caution must be taken when handling any product immersed in liquid nitrogen because of the severe burns that can occur from contact with skin. The ability to store semen for long periods generally outweighs these concerns and with experience there is little problem in handling the straws of semen.

Insemination can be done by you or by a technician employed by the bull stud to sell semen and inseminate

Once the insemination gun is loaded it can be passed through the vagina and cervix, depositing the semen in the uterus. Conception rates will generally rise as you gain more experience breeding your cows. For best results, make sure the cows are securely restrained during insemination.

Providing a clean, dry area for cows during calving reduces the risk of post-calving uterine infections and will be a more pleasant place to assist births.

The front feet of a calf generally appear first, followed by the head and then the rest of its body. Abnormal birth positions should be identified early in the birthing process so assistance can quickly be given.

cows of all and any breeds. Most studs offer training schools where you can learn the proper techniques for handling semen and inseminating cows. Information on these schools is generally available from any bull stud.

Adequate facilities for restraining the animal to be bred are essential. Chutes, breeding boxes, and other ways of keeping the animal stationary during the insemination process are required in order to successfully deposit semen into the uterus of the female.

Farmers may use a bull to breed their cows when they have little time for heat detection. Bulls can breed a female without the use of restraints or special facilities for handling the females.

Pregnancy rates may be higher using a bull, but only if that bull is fertile. That is one of the drawbacks of using a bull. If you purchase a bull, you should have the motility of his sperm checked for a satisfactory level as part of the final purchase agreement. The time lost from nonconception is financially devastating whether you are considering beef or dairy cows.

Safety is a major consideration when keeping a bull in the pasture, corral, or pen. The masculine nature of a bull, young or old, is not to be taken lightly. Failure to understand the strength and quickness of a bull as small as 500 pounds or as large as 1,500 pounds or more can lead to serious physical injury if the person or family member handling the cattle becomes the object of that bull's anger.

Seemingly calm bulls can attack at a moment's notice, particularly if they sense that someone is intruding into their territory of females. Never enter a pen or pasture where a bull is present without proper safety precautions. There are ways to restrain bulls for use within your herd and a carefully thought out plan to handle a bull should be part of your program prior to a bull's arrival on your farm.

SCHEDULED CALVINGS

Set a time of year you wish to have your cows calve. Do this very early in the process of acquiring cattle. Some farmers prefer calving in the spring to make use of the spring and summer growing season for grasses. Feed availability is greatest at these times and feed costs are at their lowest. Young calves can grow from their mother's milk and learn to graze on pasture, which is a good source of nutrition.

Some farmers prefer a fall calving schedule because the calves can grow on milk and not be dependent upon dry hay. By the time spring grass arrives, the calves' rumens have developed sufficiently to more readily utilize the forages.

Other farmers spread the calvings out during the spring, summer, and fall in order to spread out the eventual sale of market animals throughout the year. Whichever schedule seems best for you, try to eliminate winter calvings in most northern climates because of the problems associated with cold weather. Some calves freeze to death because after birth they are wet and the cow doesn't have the chance to get her calf dry in a timely fashion. There may be exceptions resulting from cows settling outside the breeding parameters you've set but, as a rule, it is best to avoid winter calvings unless you have a warm enclosed area for the cows to give birth.

A bull in the pasture with cows is one way to ensure that any cow that did not initially settle are bred.

CALVING TIME

The end result of a pregnancy is a calf. There are several considerations for achieving a high percentage of live births. First, you may choose a breed that typically produces calves with smaller birth weights. This helps ensure that the calf is small enough to pass through the birth canal of the mother and will require little or no assistance to enter the world alive.

A common-sense approach to calving goes a long way in ensuring low mortality rates. If cows have access to a dry lot or pasture, nature can do the rest. A clean, dry area for calving is ideal because it offers the mother and the newborn freedom from dirt, mud, and bacteria.

Calving areas or pens allow you to keep an eye on the whole birthing process. You may have taken care in choosing a breed with small birth weights and you may have used bulls known to sire small calves. In reality, you may have done everything possible to give the cow all that is required for calving easily. But, things can go wrong at calving time.

Most of the problems that can occur during the birthing process are caused by calves that do not exit the birth canal in the proper way, sometimes due to abnormal presentations of the calf. For various reasons, the birth may become difficult and having the cow close at hand will help in correcting the situation sooner and easier.

There are signs you can look for in a pregnant cow that will indicate when calving is close at hand. The rapid expansion of the udder of the cow or the full development of an udder in a young heifer are early signs of calving. As the time gets closer for the calf to be born, the broad ligaments, as observed by a marked sinking away on either side of the tailhead, are an indication of imminent calving. When the udder is full and the vulva is flabby and large, the cow will usually freshen within the next 12 to 24 hours.

The process of calving normally occurs in several stages. The first step is the opening of the cervix, which usually takes three to five hours. At this point the cow will start to strain and as this straining pressure increases, the placenta will break and release the fluid surrounding the calf.

A bred yearling heifer will show immature udder tissue beneath her pelvis. This is a good visual sign that she is pregnant if no pregnancy examination has been made.

During a normal calving, the front feet and nose of the calf appear first as it is forced out of the vulva. As the uterus contracts, it pushes the calf out further until the head and shoulders begin to emerge.

At this point assistance can be given if it looks like the calf is larger than expected. This can be done by gently tugging on the forelegs of the calf with a rope or obstetric chain attached to one or both legs above the ankles. If you give assistance it is better to pull only when the cow strains and pushes. Once the feet, head, and shoulders emerge, the rest of the calf will usually follow fairly easily.

A critical time in the whole birthing process is just before the hips of the calf emerge. Just prior to this moment, the umbilical cord is still attached and the mother is still assisting the calf in getting oxygen. Once the umbilical cord is torn, the calf is on its own and must begin breathing by itself in order to survive.

Generally, the shock of hitting the ground or the cold air seems to assist the calf in starting to breathe. If it does not, you may have to check the calf's airway to clear any mucous. Sometimes slapping the calf or pushing on its ribcage can jump start a calf hesitant to start breathing. If the airway doesn't seem to clear, you may need to lift the calf vertically by its hind legs to let mucous and fluid drain out.

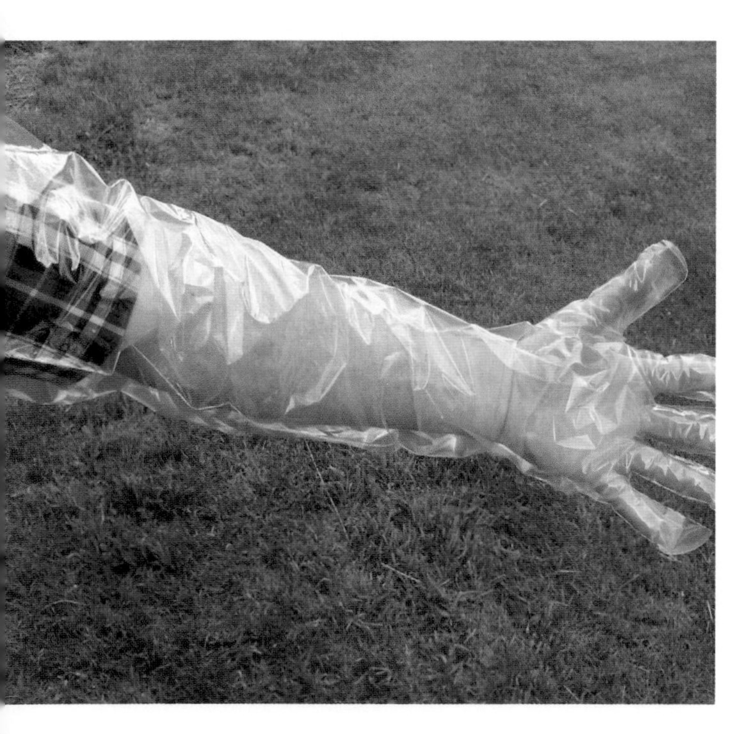

Shoulder-length disposable plastic sleeves easily slip over your hand and arm to protect your skin during the artificial insemination of the cows or heifers.

ABNORMAL PRESENTATIONS

There may be times when an abnormal presentation of the calf occurs and steps must be taken to reposition the calf while it is still inside the cow. There are numerous abnormal calf positions but the six most common ones are:

Simple leg flex in head-first position: This is when one front foot is forward in the correct position and the other is flexed back and inward.

Hind feet first: This is a backward calf. Many calves can be saved from a breech position with correct assistance.

Head-first position with head and neck turned back: The front feet are exposed but the head is turned back toward the inside of the cow.

Hock flex-hind feet first: The backward calf with the back legs turned downward.

Head-first position with rear legs under the body: The whole calf trying to come out all at once.

Backward and upside down: This is one of the most difficult situations to correct but it is still possible to deliver a live calf.

Each of these situations requires immediate attention. As the uterus continues to contract to expel the calf, it puts more pressure on the calf and the amount of room to correct these problems decreases making it more difficult to successfully correct the situation.

Prior to entering the cow's uterus with your hands and arms, thoroughly clean your arms and hands and the vulva with soap and water and use plenty of lubrication. Good sanitation at calving time will minimize reproductive problems later on.

If you feel uncomfortable trying these techniques, assistance will be required from a veterinarian or someone who has experience in assisting with difficult births. But with practice, you can learn how to successfully assist abnormal births.

Fortunately, you will not need to assist cows at calving time very often, but when things go wrong you will need some understanding of how to correct the problem and the confidence to increase your chances of saving the calf.

CHAPTER 12

NUTRIENT MANAGEMENT— HANDLING MANURE

Handling and utilizing the manure produced by the animals on your farm in the most effective way is called nutrient management. Manure is a necessary byproduct of cattle. How it is used once it is expelled from the animal will largely determine if it is a valuable asset to your farm or a costly liability.

The purpose of a nutrient management program is to use the manure components in a way that provides food for your soil through balanced applications across your fields, while minimizing any detrimental effects on the surrounding environment, such as waterways, streams, or other water sources, including wells.

With an increase in the number of farms with a large animal population, nutrient management has become a priority issue with farmers, environmentalists, government agencies assigned with the task of maintaining clean water resources, and the public at large.

Many problems encountered with manure result from runoff of these highly concentrated nutrients when the manure is spread on the top of fields. Most often

A tractor-drawn manure spreader can spread solid manure in even patterns across a field. Even spreading assures that a wide area of the soil is covered. By calculating the weight of each load, the approximate total nutrient value can be determined when finished. Solid manure is an excellent source of fiber and organic matter to help replenish the soil and loosen the particles.

Manure piled on solid surfaces, such as cement, minimizes runoff and leaching before being spread. Having the flexibility of storing manure during times when it is difficult to get into a field will assist in planning a spreading schedule and avoid times of high runoff from rain or snowmelt.

Compost piles are an excellent method of handling stored solid manure in an environmentally sound way. Periodically stirring the compost pile will help transform the manure into more stable nutrients. In turn, this can be sold as mulch, garden manures, or other forms of fertilizer.

runoff occurs in connection with frozen ground where the manure has little or no chance of being absorbed into the ground prior to rain or snow melt. Farm waste runoff into streams, creeks, rivers, lakes, and other water bodies has become one of the most contentious issues between rural landowners and urban populations that view these waterways as recreational areas. However, this issue does not have to become a problem. Well-planned usage of animal waste products—manure—can avert problems before they occur and can be a valuable asset instead of a liability to your farm.

MANURE USES

Manure is fecal material and urine expelled by an animal's gastrointestinal tract, the result of normal digestion processes. It quickly decomposes under warm, moist soil conditions and releases nitrogen, phophorus, potassium, and other nutrients into the soil.

While recognized very early in human civilization as having beneficial effects on crops and plants, manure also has had various uses not related to soil fertility including fuel, shelter construction, and sport. With the introduction of commercial fertilizers, manure became a problem to be disposed of rather than an asset to be utilized. Today the emphasis is on keeping manure runoff from reaching water sources and using it to help improve the soil fertility, structure, and composition because of the nutrients manure contains. Field plants absorb nutrients from the soil in order to grow. Replenishing these nutrients with manure has been a solution to lowering outside fertilizer purchases.

Solid manure, when mixed with bedding such as straw or hay, can add fiber and organic matter back to the soil. This combination, when plowed or stirred back into the field, can loosen soil particles and allows the soil to become more porous, absorb more moisture, and create more capacity for holding water. Loosening the soil particles relieves compaction of the soil that will provide the plants with a soil structure that allows a better root system to develop for better growth.

CALCULATING MANURE QUANTITIES

The amount of manure produced on your farm can be calculated to a fairly close quantity. Typically, a mature beef animal can produce 75 pounds of urine and fecal material per day. A dairy cow can produce around 100 to 115 pounds per day. Multiply that by the number of mature animals on your farm and you have a total for the daily volume of manure produced. The type and size of your beef or dairy herd determines the quantity and types of manure to be handled.

All these nutrients in manure, beneficial as they may be, can be considered pollutants if they enter the streams or groundwater systems. Improper manure management can have a serious effect on wells and the quality of drinking water. It is essential to understand the proper handling of various types of manure produced on your farm to avoid potential causes of water pollution.

SOURCES OF MANURE POLLUTION

Sources that can contribute to water pollution problems resulting from improper manure management include poor land application techniques, runoff, leaching, open lots, and the improper location of manure storage facilities.

Manure application to your fields and pastures needs to be controlled in a way that does not allow the nutrients to seep into water systems. This can occur if manure is applied to frozen or snow-covered ground, to saturated soils, to sloping fields, or to areas that are too near to streams or ditches, or by an excessive application rate.

Runoff of manure content can occur when rain or snow mixes with the manure resting on the ground or an open lot where the cattle are fed or congregate. Heavily used areas generally have little or no grass coverage to filter out these materials before they drain to a water source. Runoff to streams is more likely when the field selected for spreading the manure is located next to a stream, pond, lake, or other surface water rather than when the manure and water source are separated by another field, pasture, or grass buffer strip.

Leaching becomes a problem when the application on a field or pasture becomes too concentrated and the leaching of the nutrients into groundwater sources is increased. The soil condition and type will largely determine the amount of nutrients that can be adequately filtered before

Medium- and large-sized beef farms use runoff pits as collection basins for manure, urine, and rainwater that runs through cattle areas to prevent it from entering roadside ditches, streams, or farm valleys.

Many large dairy farms use cement-lined manure pits to store liquid manure until fields are available for spreading the manure. Lining these pits with impermeable materials prevents leaching and makes them easier to pump out. Safety precautions need to be taken, such as perimeter fences and locked gates, to keep children and animals from accidentally falling into the pits.

it reaches the groundwater supply. Lighter sandy soils will not filter as many nutrients as heavier loam and clay soils.

Open lots are areas where cattle congregate in large numbers to eat, drink water, or lie down. This area has the greatest potential for runoff because of the high concentration of cattle in a small area with little constraint of the water from a heavy rain or melting snow. Typically these areas are close to buildings and the volume of water added to this area from rain runoff can add substantially to the amount of water draining through the lot.

Location of manure storage facilities can also be a potential source of pollution. The bacteria, nitrates, and other substances found in stagnant manure piles have the potential for well contamination if they are located too close to the water supply system.

MANURE HANDLING

The manure produced on your farm can be sorted into three types: solid, semisolid, and liquid. Each of these types needs to be handled in a different way to make effective use of them.

Solid manure consists of a combination of fecal matter and urine that has been mixed along with dry bedding materials, such as straw, hay, sawdust, wood shavings, corn fodder, or any other material that has become part of the bedding. Typically, solid manure is handled with a manure spreader that applies the material on croplands to be used as a fertilizer.

Semisolid manure contains little or no bedding materials and has a consistency between liquid and solid, much like a thick pudding or slurry. This consistency creates some difficulty in storing it for spreading onto the fields. One solution is to mix in bedding materials to get a firmer consistency or to use a sloping floor to allow the liquid to drain to a holding area and then storing the solids in another area. By separating the parts, semisolid manure becomes easier to handle.

Liquid manure contains no bedding materials and is typically a combination of urine and feces. This requires a storage facility that does not let any of the nutrients leach into the ground, such as cement-lined pits or upright structures. In confined buildings there is a slotted floor system where the manure drops down into a pit. This storage system requires pumping the manure from the pits into tanks that can be taken to fields and spread on the open ground or injected into the soil.

MANURE STORAGE OPTIONS

There are several options available for storing the manure produced on your farm. Farms with greater numbers of beef or dairy cattle need larger storage capacities. With smaller numbers of animals, storage options may be more affordable because the startup costs can be minimal.

Composting is one way to handle manure in an environmentally sound manner and can be a potential revenue source for your farm. Composting is the active microbial treatment of solid manure by using oxygen as the main catalyst for this process. The organic matter is allowed to decay in a pile or windrow. Decaying organic matter creates heat and a compost pile of manure, depending on its density, can reach temperatures of more than 160 degrees Fahrenheit at its core. Oxygen is required for composting so the pile needs to be turned or stirred occasionally for the material at the edges of the pile or windrow to become part of the heating process. A tractor with loader or other equipment designed for stirring piles or windrows can accomplish this task.

In pasture programs, the manure is spread in paddocks as the cattle move while grazing. With a good pasture rotation, the fields can be fertilized without a lot of mechanical assistance.

Properly designed buffer strips along creeks and waterways provide a filter to prevent pasture or field runoff from entering streams. These buffers may be grazed for short periods of time to make use of the available grasses lining the edges.

Composting has the advantage of reducing the volume of manure and transforming it into a more stable nutrient form. These nutrients, when spread on the fields, are slowly released into the soil for crop nourishment. Because of this, the nutrients in composted manure are less likely to be transported off the site through runoff and leaching into the ground water.

A second advantage of composting is that manure can be stored until weather and field conditions are better for hauling and the absorption rate of the nutrients by the soil is greater. Composted manure can also be sold as an off-farm fertilizer, soil additive, or mulch. Garden stores offer these products in plastic bags for gardeners, vegetable growers, and others. Because the manure has been broken down into less volatile nutrients, it provides a more benign product to be used by the general public. The development of a market for composted manure may have more to do with distribution than with actual production. Information regarding this market and how to develop it is generally available from your county agriculture extension office.

MANURE HANDLING ON PASTURES

One overlooked source of manure runoff control is pastures. The most economical way of spreading the majority of the manure produced by cattle is to let them distribute it themselves in the pastures while they graze.

If there is enough pasture to feed the animals and still sustain the vegetation, the solid manure produced by each animal is more or less spread uniformly around the pasture as they move about and requires no extra handling on your part. Any possible runoff effects from pastures can be minimized by rotating pasture areas, avoiding steep slopes, streams, and drainage ways. The manure the animals create will slowly decompose in the pasture and provide a habitat for a wide range of useful insects that help break down the manure into nutrients used by the soil.

MANURE MANAGEMENT PRACTICES

Nutrient or manure management practices are required for farms expanding beyond a certain threshold for animal

Cattle will seek water to cool themselves during hot weather. Building buffer strips prevents cattle from accessing the streams or waterways and prevents pollution.

numbers or those involved with government farm programs. Nutrient management plans are developed with professional experience and approval. These plans develop a program for each farm where the nutrients produced from manure are accounted for in the total field application of fertilizers, whether purchased or not.

Accounting for nutrient application on all fields ensures that excessive amounts are not used, thereby limiting the possibility of leaching or runoff into the ground water supply. Areas of high vulnerability for runoff are identified and the spreading of manure in those areas becomes restricted. By not exceeding crop requirements, the soil is not saturated with nutrients it cannot absorb.

Soil tests are one part of a good manure management plan and are the starting point for determining the nutrient content of your fields. The results explain the nutrient requirements of each field and will serve as a guide to application rates.

Another component of a good nutrient management plan is assessing crop nutrients already available in fields, such as manure, legumes, and organic wastes. With a calculation of the amounts already present, the total available nutrients can be deducted from the soil test recommendations. This will help determine what, if any, additional fertilizer purchases are needed.

Fresh manure pats will slowly dry in the sunshine and their thick consistency does not make them as vulnerable to runoff from pastures or fields as liquid manures.

SOIL CONSERVATION PLAN

If you plan to participate in any federal farm programs, a soil conservation plan is required. The conservation plan is a part of any nutrient management program because it identifies crop rotations, the slopes of all fields, and the conservation measures you will need to follow to stay within the tolerable limits of soil erosion.

This plan also identifies which fields may have restrictions for spreading manure because of a close proximity to waterways, especially in the winter. One component of this plan includes identifying the best time of year to spread manure. This will depend upon the manure-handling system on your farm. A farm with manure storage has a plan different from one that requires daily manure hauling. Many counties require a nutrient plan for any farm constructing new manure storage facilities or expanding their livestock operations.

Properly designed buffer strips along stream banks adjacent to fields with a potential for runoff can reduce the amount of manure entering a stream. These grassy strips

Manure pats are spread in the field as the cows walk about. They decompose and provide nutrients for the plants and homes for insects that feed on the fibers and organic materials. Insects provide food for grassland birds. Grass will grow through decomposing manure.

Liquid manure can be safely hauled to fields in large tanks filled at the top. Once in the field, injector knives set into the soil deposit the manure below the surface and prevent runoff.

help stop, filter, and hold back the sediments of runoff. Fencing plays an importanr roll in buffer strips because it keeps cattle from overgrazing banks and waterways.

PLAN AHEAD

Many problems with farm manure can be avoided with a plan that involves the best utilization of it over the entire farm, while staying away from potential runoff into waterways. Nutrient and manure management plans and a conservation plan for your farm can be developed with help from your county agriculture extension agent or the Natural Resources Conservation Service.

Streams and waterways are vulnerable to manure runoff any time of year. Protecting them with good fencing and buffer strips keeps potential pollution at a minimum.

CHAPTER 13

HUSBANDRY— GENERAL DAILY BEEF CATTLE CARE

Cattle husbandry is using the resources you have at hand to provide your animals with a safe, comfortable, and fulfilling existence. It acknowledges that cattle need day-to-day care and attention, even though they may be self-sufficient in many ways. Animals cannot do everything for themselves and you will need to provide care for them including giving vaccinations, observing their physical condition on a daily basis, and helping them avoid problems that are often unseen until they happen, such as bloat.

The best cattle producers try to anticipate problems and take steps to eliminate potential causes as best they can. For example, a good farmer would recognize that a simple thing such as a fence wire being down could cause young animals to investigate the possibilities of going through the gap to see what is on the other side, which could cause a snowballing effect of half the herd ending up in your neighbor's field. Additionally, health concerns, such as illnesses, lameness, or other physical ailments, will need your attention.

Some potential problems requires daily observation of your cattle; perhaps several times a day if the weather

Cattle are social creatures and will seek the company of other cattle. Calves display this behavior at a very young age as they confront each other in the pastures and fields.

Observe your cattle when putting them in the pasture for the first time in the spring. Feeding dry hay before they go on grass will diminish the risk of bloat.

Antibiotics have played a major role in treating cattle illnesses for over 50 years. As certain types of bacteria become resistant to an antibiotic, new strains must be created to stay ahead of this mutating resistance. Antibiotics, while useful, have become a mixed blessing. They must be properly stored and refrigerated so that they are available for use when necessary.

is extreme. Problems can surface quickly and your response to these events may determine whether an animal lives or dies.

Raising cattle is a business, and like most successful businesses, managers who pay attention to the minute details often survive the difficult times more readily than those who don't.

As a cattle owner you have a responsibility to provide your animals with conditions where they can grow from quality feeds, reside without fear, and be treated in a gentle and humane manner. These are similar circumstances to what you would provide for your children. In some respects the more you work with your cattle, the more your cattle become an extension of the family. With a herd of 2, 10, or 30 cows, it is likely several will seek your attention rather than the other way around.

At first this may seem an odd consequence of raising cattle but it is in your best interests, financially and philosophically, to have animals that are content and unafraid. Docile animals will grow faster and be safer for you and your family to work with.

UNDERSTANDING CATTLE BEHAVIOR

Animals respond to the treatment they receive. Cattle, in particular, will respond in a negative way to situations where they feel threatened. This could be a simple thing, such as being the last one away from the water tank and suddenly realizing the rest of the herd walked back to the pasture unnoticed by the lone straggler. Eliminating the obstacles that may trigger fear in cattle requires some anticipation on your part. Not all situations will be perceived with fear by the animal, but removing those that might be is a safe and humane thing to do.

Cattle have 360-degree vision, which means they are alert to all kinds of movements. Dark shadows where they cannot distinguish between certain objects can lead to situations where they respond differently than in a well-lit area. During evenings or the winter months when there is less light, developing a familiar routine for your animals will help avoid surprises.

Cattle are creatures of habit and their familiarity with an area will allow them to respond in ways they feel comfortable. When they are moved to a different area, such as a new pasture, field, or corral, they will first investigate the area by walking around the perimeter fence. It may seem that they are looking for a new way out of the field, but they are simply familiarizing themselves with their new surroundings. Once satisfied, they will begin to cross the pasture and start eating. Their fear has been removed as they feel comfortable with the change.

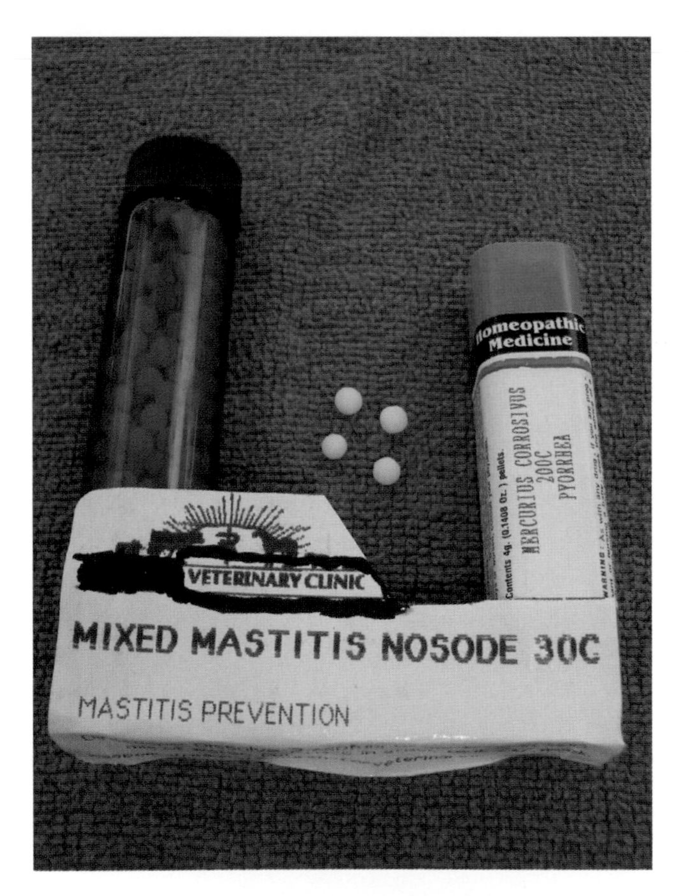

Homeopathy is an alternative treatment protocol that allows the animal's immune system to fight off illnesses. Homeopathic remedies are given as small pellets and work to elicit specific immune responses. The animal's body works to fight off the effects rather than suppress the symptoms.

SOCIAL CREATURES

Cattle are social creatures, and by watching a group of young calves out in a pasture this becomes quite obvious. While the cows may not mingle as much, young calves seem to seek out each other. This is learned behavior because the calves, soon after they are born, have more interest in themselves and finding their mothers for something to eat than anything else. After they have grown enough to stray from their mothers, they seek out other small calves.

It can be fascinating watching young calves develop their socialization behaviors. It is very much like watching a young child trying to befriend another child on a playground where there are touches, hugs, and a little animosity toward the other person. Calves display a similar behavior in the pasture. They will sniff noses, butt heads, and run along with each other. Eventually, the young calves will find a place where they can lie down together away from their mothers. It's not that their mothers don't keep an eye on them, because they do, but it is usually from a distance. Their mothers seem to instinctively know that it is good for the calves to be together and to be around each other. In these situations one mother often serves as guardian of all the calves while the other cows go off in a different direction. Mothers take turns watching the calves from a distance and stand guard against any threat.

Cows have a social order that is quickly established when they are brought together from outside sources. Bringing new animals into a herd requires some forethought as to their size and number. A single animal brought into an existing social order will quickly be placed in an adversarial relationship with other cows. In many cases this can't be helped because this will occur at some time during their initial introduction and there will be little that you can do about it.

It is not in the nature of cattle to continually fight each other. Constant fighting is not normal, although some dominant cows can make it uncomfortable for those lower

Healthy animals can withstand different climate conditions. Being alert to the needs of your cattle may require several daily observations and recognizing problems early may avoid long-term effects.

When a female calf has been vaccinated for brucellosis, the attending veterinarian will attach an official vaccination tag to one ear as well as tattoo the month and year on the inside of the ear. Properly done, the tattoo will confirm the official vaccination status should the ear tag be lost.

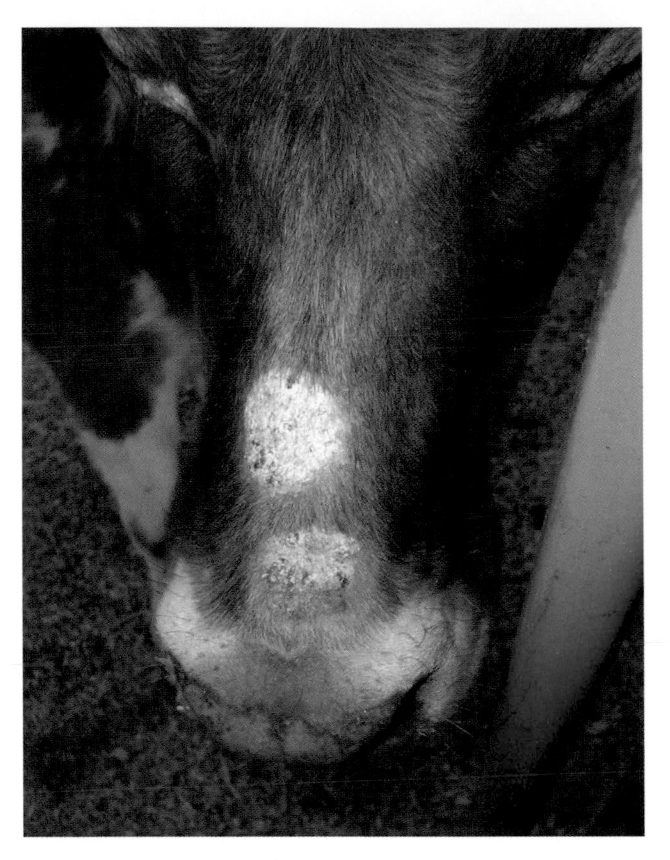

Ringworm is a highly contagious skin infection that affects young animals. While it looks tender, there is little pain involved with ringworm, although the animal will rub it to relieve the itching. Be sure to use good sanitation practices when treating ringworm.

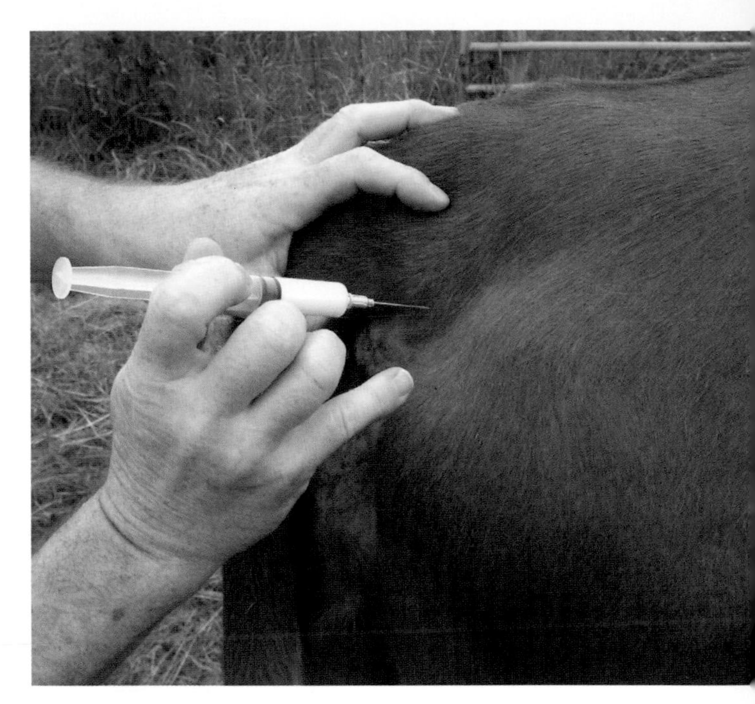

One acceptable injection site on an animal is the area between the tail head and the pin bones. This eliminates the need for muscle injections, which lower carcass quality.

on the social order. The realignment of the social order will be quickly established and the cattle will soon resume their activities as if nothing happened.

There is growing evidence that some social behaviors in cattle can be affected by their contact with humans. The more contact there is when they are young, even as day-old calves, the less fear they have of humans. This lack of fear quickly translates into calmer animals when you are moving about them, whether it is in the pasture, farmyard, or fields.

HERD HEALTH MANAGEMENT

One rule that can define any cattle grower's experience is that animals can get sick even with the best of care. The key is to minimize the number and severity of those illnesses to the greatest extent possible.

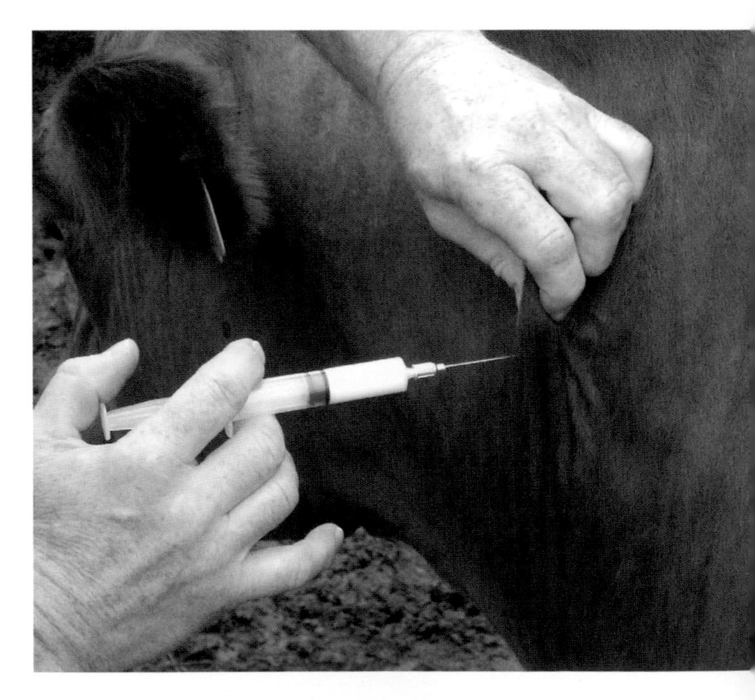

The area around the neck where few muscles are located is an acceptable injection site. Intramuscular or subcutaneous injections should be given in this area as marks left from needle punctures are not as likely to cause a discounted carcass.

Various forms of animal identification can be used to mark your cattle. These include ear tags, neck chains, freeze brands, photographs, and any other item that can help you distinguish one animal from the other.

The first step in minimizing the effects of any sickness is to visually inspect the animals on a daily basis. This does not necessarily require a great amount of time, depending on your herd size, but walking out to where the cattle are and looking for signs of any health or physical problems will help keep you abreast of their condition. You will be able to quickly identify animals that may be listless, limping, have droopy ears, or other signs that they may not be eating normally, which can indicate a problem.

Identifying such problems will allow you to move quickly to assist them. Generally, an animal that exhibits symptoms of illness will not overcome it without your intervention or help. You become their diagnostician, doctor, and pharmacist all at once.

TREATING ILLNESSES

You can treat most cattle illnesses without any specialized education. Experience is the best teacher, but you can gain insight and understanding by reading about cattle health and the practical application of administering treatments.

There will be times when you have no choice but to call a veterinarian to help with a particular health issue.

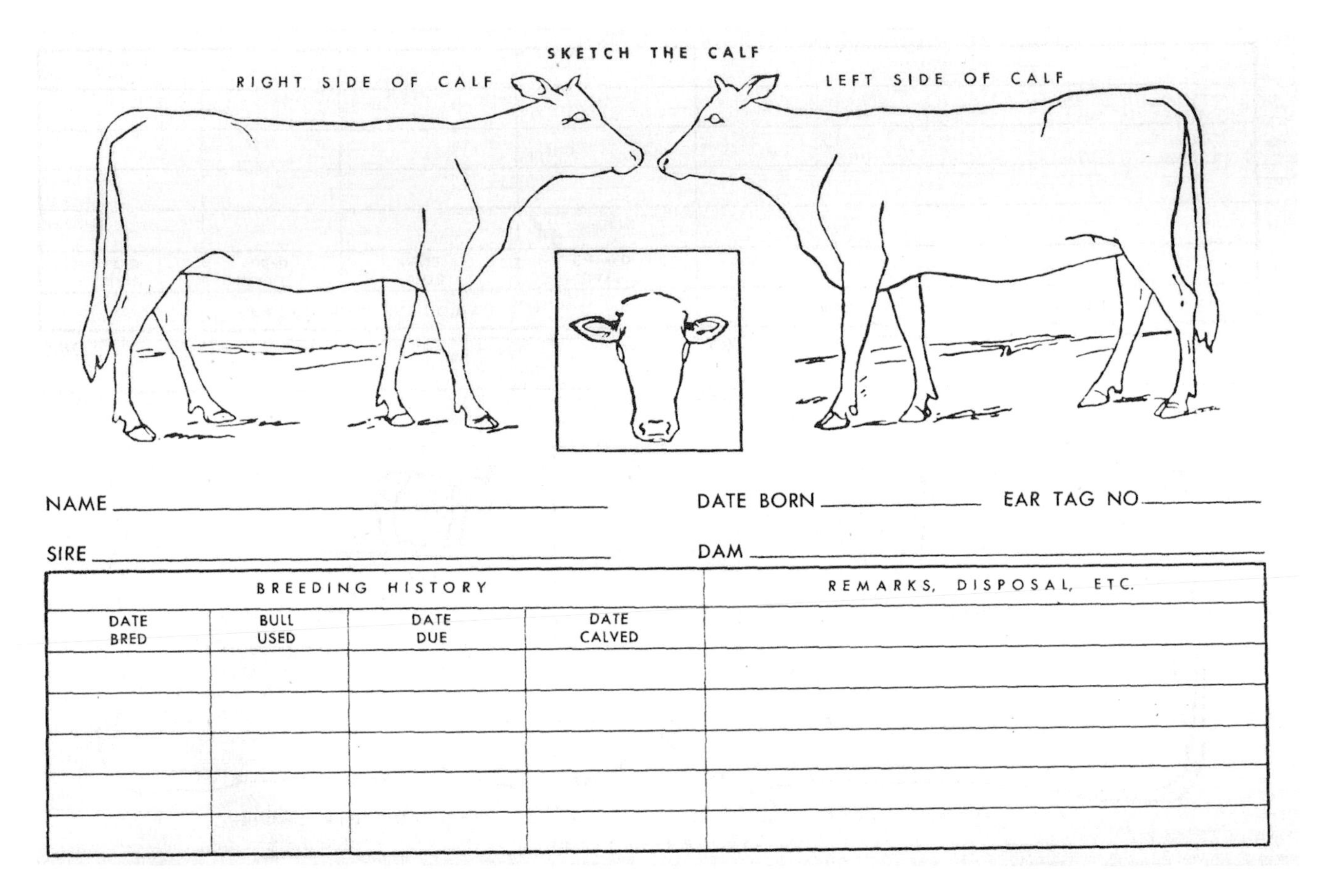

SKETCH THE CALF

RIGHT SIDE OF CALF

LEFT SIDE OF CALF

NAME ______________________ DATE BORN __________ EAR TAG NO __________

SIRE ______________________ DAM ______________________

BREEDING HISTORY				REMARKS, DISPOSAL, ETC.
DATE BRED	BULL USED	DATE DUE	DATE CALVED	

This is an example of a breeding record that can stay with your animal for its entire life.

However, the severity and extent of the problem may have to be solved by your own initiative, such as during a case of bloat where time for resolving the problem is extremely short and a call to a veterinarian most likely will be too late. A licensed veterinarian is required to administer vaccination injections, medical euthanasia, and certain antibiotics or steroids.

CONVENTIONAL TREATMENT

Conventional treatment of cattle generally involves the use of antibiotics to relieve the symptoms of the illness or disease. While antibiotics may correct the problem in the short run, residual effects may affect marketing those treated animals.

The use of antibiotics to treat disease in food-producing animals started in the mid-1940s, and by the early 1950s the industry saw the introduction of antibiotics in commercial feed for cattle. In the past 40 years, antibiotics have served three purposes: as a therapy to treat an identified illness; as a prophylaxis to prevent illness in advance; and as a performance enhancement to increase feed conversion, growth rate, or lean meat yield.

Bacterial diseases generally cause pain and distress in an animal, as well as an economic loss if unchecked. Antibiotics can be used to help reduce this suffering and distress and speed the recovery of an infected animal.

When used responsibly, antibiotics can be an essential element in the fight against animal diseases. In rare instances they may be used to prevent diseases that might occur in a herd or group of animals if a high probability exists for most or all the animals becoming infected.

However, there are some cautions when using antibiotics, especially on a routine basis. In animals, as in humans, a significant proportion of those treated for infectious disease would recover without antibiotics. One study has shown that every year, 25 million pounds of antibiotics, roughly 70 percent of the total antibiotic production in the United States, are fed to chickens, pigs, and cows for non-therapeutic purposes, such as growth promotion. This report also showed that the quantities of antibiotics used in animal agriculture dwarf those used in human medicine. Nontherapeutic livestock use accounts for eight times more antibiotics than human medicine, which uses about three million pounds per year.

The resistance of bacteria to drugs, such as antibiotics, has been documented in several studies. New antibiotics need to be manufactured as bacteria mutate and become resistant to the previous generation of drugs. The more antibiotics are used, the greater chances of a residue entering the food chain that can affect the general population.

The excessive use of antibiotics can produce drug-resistant bacteria, which may become difficult or impossible to treat. Any animal that has received antibiotic treatment must be withheld from the market for a specified time and the withholding times must be followed by the producer or veterinarian.

This doesn't mean that antibiotics shouldn't be used. But if they are used it is imperative that they be used judiciously

Neck chains with tag numbers are simple forms of identification. Once properly fastened, a neck chain will stay on for as long as the cow is in your herd.

For young animals, grazing is a learned behavior that quickly becomes part of their lives. Older animals will quickly adapt to grazing once put on a pasture.

Activities offered through 4-H and FFA chapters across the United States provide opportunities for youth to experience the pleasures and challenges of showing beef and dairy cattle in organized competition.

Showing cattle develops a variety of skills that can be useful throughout life. A cattle project involves feeding, grooming, halter training, and exhibiting an animal at a county or state fair or a breed-sponsored or -sanctioned show. Depending on the quality of the animal or the interest level of the member, regional and national shows also provide excellent learning opportunities on a broader scale.

While prizes, premiums, and recognition are part of the benefits of showing, the chance to work with an animal over an extended period of time has many residual, often intangible, benefits for the 4-H or FFA member.

Besides becoming acquainted with other 4-H or FFA members from their county or state, young men and women learn a variety of skills not readily attained in other ways. These skills range from the daily handling and care of animals to management or ownership of animals, which can vary in age from young calves to cows. Whatever the age of the animal shown, handling and caring for it involves decision making, discipline, and patience, which can be useful in careers later on.

Managerial or Ownership Projects

A managerial project allows a 4-H or FFA member to raise, care for, and exhibit an animal without having to own it. This generally involves members whose parents, neighbors, or friends own the animal and allows the member to feed, train, and raise it for a specified period of time. This arrangement allows the member to learn how to properly care for an animal without the purchase cost. In this case, the member is generally responsible for all expenses during the length of the project including feed, veterinary care, and insurance. In return, the member receives all awards or financial premiums won at any shows where the animal is exhibited.

Written agreements for managerial projects are available from county 4-H extension offices. These agreements outline the responsibilities of the member and the sponsor or owner of the animal and are a good introduction to understanding the requirements for completing a successful project.

An ownership project involves a member owning an animal outright and receiving all benefits from the project year. The member also assumes all costs and expenses during the term of the project.

Developing Life Skills

Cattle projects help young 4-H and FFA members develop recordkeeping and budgeting skills. Because they assume control of the animal, members need to keep records of the costs, expenses, income, and health of the animal in order to complete a project report at the end of the year. 4-H and FFA programs provide training in keeping accurate records and serve as a model for real life experiences.

This is perhaps the most important aspect of showing cattle, whether they are beef or dairy and whether they are young calves or cows. Showing can be fun, but the lessons young men and women can learn from raising a calf into a cow over several years are beyond financial measurement.

Training an animal to lead by halter, which is required at shows and fairs, involves discipline and patience by the member and communication with an animal. The latter involves unspoken and verbal commands between the member and the animal and the use of hands on a halter to signal nonverbal direction. All are intended to show an animal to its best advantage.

Discipline involves the daily dedication to animal care and scheduling the time to provide proper nutrition, water, and housing to make sure an animal is comfortable and well fed. This ensures adequate growth so the animal is comparable in size to others within its own age group when shown.

Patience is required when training an animal to lead by a halter so that it understands the commands given to it by the member. An animal cannot be expected to reason the same as a person. Steady, consistent commands will lead to less confusion on the animal's part when it is learning what is required for showing performance.

Personal Growth

Participating in shows allows the member to learn about the ethics of raising and showing animals in competition. Ethics are an important aspect of showing and help keep the competition fair for all exhibitors.

Rules and codes of ethics have been developed by the beef and dairy breeds regarding practices that are allowed and those that are discouraged or not tolerated. Basically, any human manipulation that alters the physical appearance of the animal, aside from the hair coat, is prohibited.

The reasons for this rule should be obvious. The alteration of the animal's physical structure, whether it is the bone or muscle, is detrimental to the well-being and health of the animal and cannot be tolerated for the sake of winning a prize.

Learning the ethics of properly handling and raising animals is key to the emotional and mental growth of the member. Understanding the ethical choices available and learning to make them is one of the most valuable lessons young people can learn. Guidelines for understanding these ethical practices are available from the breed and industry organizations and from county 4-H extension offices.

Leadership Skills

Showing cattle provides 4-H and FFA members with an opportunity to develop leadership skills. Clubs and chapters provide adult supervision to help members learn how to develop their own self, as well as their animal projects.

While under adult guidance, members have opportunities to lead discussions. In the case of older members, they can teach younger members showing techniques and methods of raising animals that support the ethics training.

The importance of learning leadership skills, discipline, and patience at a basic level serves as a platform from which the member can build his or her personal integrity. This can have significant advantages when he or she later enter the job market. Having an understanding of animals and the ability to handle them correctly and ethically can provide members with many career choices not available to their contemporaries.

Careers in animal-related or scientific fields can be started with a project as simple as raising and showing cattle. People with experience in handling animals, ethics training, recordkeeping, and leadership abilities are always in demand by employers. More information about the 4-H program is available from county 4-H extension offices. Information about the FFA program is available from a local high school chapter or the state FFA office.

and only when warranted. In some specific cases, such as Johnne's disease, it is better for the animals to be culled rather than put through any treatment program because in most cases, the treatment is ineffective and uneconomical.

Familiarizing yourself with diseases that require antibiotic assistance will help avoid unnecessary usage. A discussion with your local large-animal veterinarian may provide other insights that will be useful in your program.

Water tanks can be filled by using wind power, such as windmills, or by a gas or diesel pump that pulls water from a well to the tank. This will make water available for pastured cattle some distance from a building where no electrical access exists for pumping water.

HOMEOPATHIC TREATMENTS

Interest in using alternative treatment programs for animals has been growing because of the concern of residual effects of antibiotics, both in animals and in the human food supply chain. Information is available for homeopathic treatment systems and this approach is ideal for those considering organic, sustainable, or biological farming methods with their cattle.

Homeopathy and herbal treatments have a place within any farm health protocol. In years past these alternatives have been shoved aside for the quick fix of antibiotics that became inexpensive and readily available. The pressures of large-scale beef production and dairy cattle confinement almost made it imperative to routinely use antibiotics to control any systemic problems. In some cases it was thought that using antibiotics routinely could replace good management practices. This does not have to be the approach used on your farm.

One of the ideas homeopathy is based upon is that bacteria are not necessarily a bad thing and that they do not need to be destroyed. It is not the illness that is being treated but the animal's reaction to it.

Homeopathy treatment involves the natural stimulation of the animal's immune system so that it can fight off the bacteria that might otherwise cause a disease. It is known that antibiotics have a suppressing effect on the animal's immune system while they are fighting bacteria. However, antibiotics indiscriminately fights both the good and bad bacteria.

Providing the animal with the ability to use its own body to fight off disease-causing bacteria benefits the animal in the long run because its physiological system is in better condition. Building up the health of the whole herd and increasing the resistance of its individual members to disease will help induce greater growth rates and milk production. A healthy cow produces healthy meat and milk products. Two approaches can be taken when using a homeopathic system on your farm: preventative and therapeutic, which is the emergency treatment of individual cases.

It is generally accepted that a program of preventive medicine is better than treatment, and homeopathy is well suited to this approach. Homeopathy can be used to support the development of a calf's immune system before it is even born by working with the pregnant mother. The first six months of a calf's life are the most important and will determine, to a large extent, its health later in life. A sickly calf does not become a healthy cow overnight but a healthy, vital calf can have the ability to stay healthy.

A homeopathic product is administered by means of a remedy. Remedies are derived from all-natural sources including animal, mineral, or plant and their preparation is made by a qualified pharmacist. A remedy is used in doses. It is usually marketed in one-gram vials and is based on a system of potencies. In tincture form the remedy is added to a sugar-granule base and allowed to soak, after which it becomes stable and can remain active for months. In some cases, if properly stored, the remedy can remain active for several years.

These vials usually contain sufficient materials for several doses, or administrations, for the animal. A dose may consist of five or six small pellets given at one time, depending on the condition being treated, and may have several applications during one day.

The dose is placed directly on the animal's tongue and dissolved by the saliva. The animal does not need to swallow the granules because homeopathic remedies can be absorbed through the palate or tongue. The potencies are determined by the dilutions made from the refined crude product. This refinement develops the inherent properties of the remedy used for the specific needs of the animal. Homeopathy is a legitimate route for treatment of animals but is not a substitute for preventive measures like good nutrition, air quality, and proper sanitation.

HERBAL TREATMENTS

A second alternative treatment is herbal, which also uses remedies based on preparations made from a single plant or a range of plants. Applications are by different routes and methods depending on the perceived cause of the disease condition.

These applications can be made from infusions, powders, pastes, and juices from fresh plant material. Topical applications can be used for skin conditions, powders can be rubbed into incisions, oral drenches can be used to treat systemic conditions, and drops can treat eyes and ears.

A refrigerated medicine chest that can hold pharmaceutical products used in your health programs should have a safety lock so only authorized persons may use the antibiotic products.

A labeled container is a safe and environmentally friendly method of disposing of needles that have been used for treatments. Alerting family members or employees of their importance may avoid inadvertent punctures on hands and fingers.

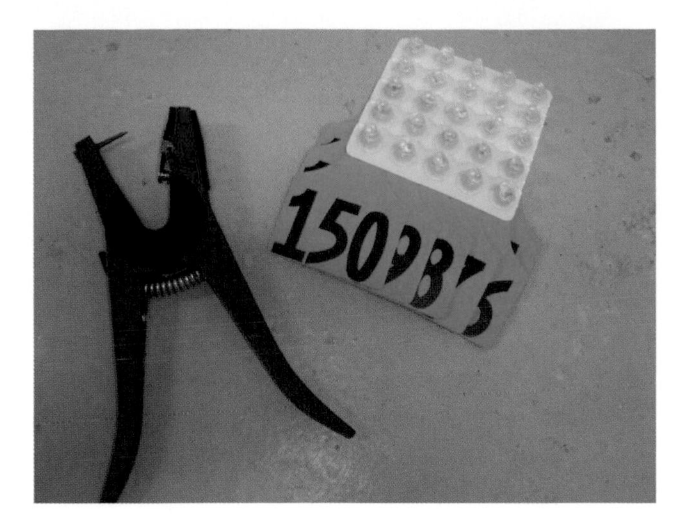

Ear tag numbers are easy to read and provide a good form of individual animal identification. An ear tag applicator makes the tagging process simple and easy when the animal is properly restrained.

Information is available from alternative stores or books published on these topics to help explain more about these products and their usefulness.

COMMON HEALTH PROBLEMS

Many health issues have the potential to affect the performance of your animals and their lives. Good planning and management, along with the use of common vaccines, will usually enable your cattle to avoid most disease problems.

It is wise to find a local veterinarian who includes cattle in his or her practice to consult about a herd health program. A list of veterinarians in your area or region is usually available from your county agriculture extension office and should be consulted prior to bringing any cattle onto your farm.

HANDLING COMMON HEALTH PROBLEMS

There are some common health problems with cattle that you can handle on your own without the assistance of a veterinarian including bloat, scours or diarrhea, pneumonia, clostridial diseases, parasites, and infections of the skin.

BLOAT

Bloat is a condition where immediate attention and action is required. Cattle spend around eight hours a day chewing their cud and belching. If that process is severely hindered, the gasses built up in the rumen are unable to escape.

Bloat is more common on pure stands of young, lush alfalfa hay rather than a mixture of grasses, but bloat can occur at almost any time of year in any kind of pasture. The fiber content of young alfalfa plants is low and protein content is high. Fiber stimulates the rumen and allows for normal fermentation and function. As the low-fiber materials ferment, a slime-froth forms on the top of the rumen contents and prevents gas from being released by belching.

As the gas pressure increases, the cow stops chewing and the rumen will swell. In the meantime, all the internal organs—the lungs, heart, liver, and intestines—are being squeezed by the enlarged and tightly pressured rumen, which makes them work harder. In some cases the death of the animal is due to a heart attack but most die from asphyxiation because the pressure has squeezed all the air from the lungs.

If caught quickly, bloat can be just an uncomfortable condition that is treated with an oral drench of vegetable oil. The oil disperses the gas bubbles and allows the normal return of belching. Sometimes a rope halter placed on the animal's head with the lead strap passed tightly through its mouth will force the animal to chew the rope. After several minutes the gasses will evacuate in much the same manner as if the cow were chewing her own cud.

In the extreme situation with very little time left for the animal's survival, you may need to slice through the hide on the bulging upper left side of the animal with a knife or puncture it with any sharp object to rapidly eliminate the gas and pressure. Later, remedial measures will need to be taken to sew up the opening and to ensure that the fermenting rumen materials don't infect the body cavity, but at least you will have saved the animal.

Bloat can be minimized by keeping cattle out of a pasture while they are very hungry or when there is frost on the plants. Feeding dry hay before they go out to pasture will ensure they are not overly hungry and the dry fiber will help them acclimate to the pasture conditions.

SCOURS

Diarrhea (scours) is a common aliment of newborn calves and young animals. The prevention of diarrhea is easier and more successful than treatment. When calves develop diarrhea, dehydration is a prime cause of death. Electrolytes and fluid given orally or intravenously every two hours is the most effective method of treatment.

However, the best way to prevent diarrhea in calves is to provide pregnant cows with adequate nutrition, especially the final two months before calving. This helps increase the calf's immunity to disease after it is born. After birth it is important that the calf receives colostrum milk in the first six hours of its life. This initial milk contains many antibodies that the calf can quickly and readily absorb through its stomach and intestine. As the first 24 hours pass, the ability to absorb antibodies decreases until it becomes negligible.

PNEUMONIA

Pneumonia in cattle usually occurs due to stress, changes in weather, and infectious agents all occurring at similar times. Pneumonia is most common in calves that have been weaned or kept in wet bedding conditions and are subject to drafts. Developing a vaccination program with your veterinarian including IBR (infectious bovine rhinotracheitis), PI3 (parainfluenza type 3), BRSV (bovine respiratory syncytial virus), and BVD (bovine virus diarrhea) will provide a broad-spectrum defense against respiratory diseases. Included with this may be five-way Leptospira, which will prevent water-borne infections related to organ dysfunction.

CLOSTRIDIAL DISEASES

Clostridial diseases are a group of related infections that may cause sudden death, especially in young cattle. These diseases include blackleg, enterotoxemia, malignant edema, and black disease. Vaccines are available to stimulate the immune system of the animal. Cattle should be vaccinated early in life with appropriate boosters given at later dates, if these diseases are a problem in the area where you farm.

PARASITES

Internal parasites may be a problem when cattle are grazed on the same pastures several years in a row. Deworming may be needed to minimize the parasite load in the body of the animal to allow proper weight gains. Products are available for use at specific times during the year and your local veterinarian can provide you with advice on how best to

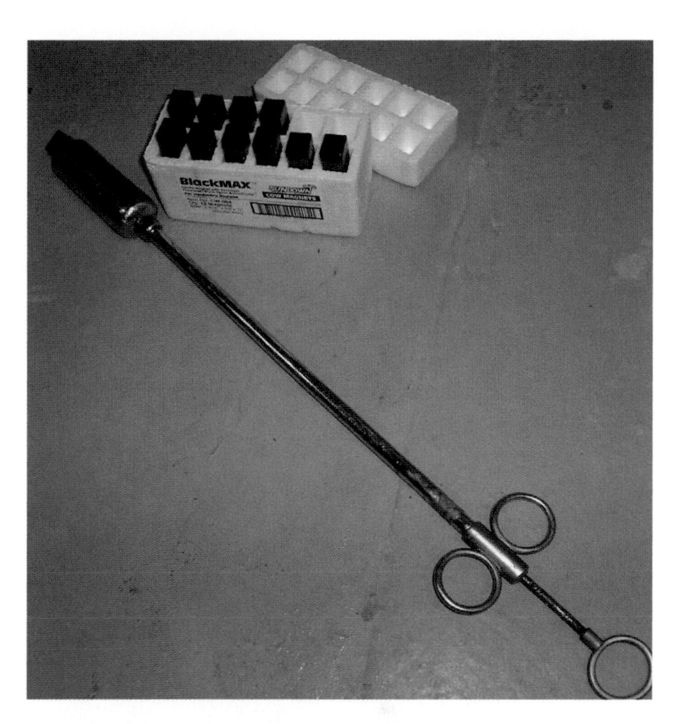

A balling gun is a long metal tube with a large hollow end for holding boluses or magnets. The gun is passed down the cow's throat and the ring plunger deposits the bolus or magnet below the cow's swallowing point to make it impossible for the cow to cough it back out.

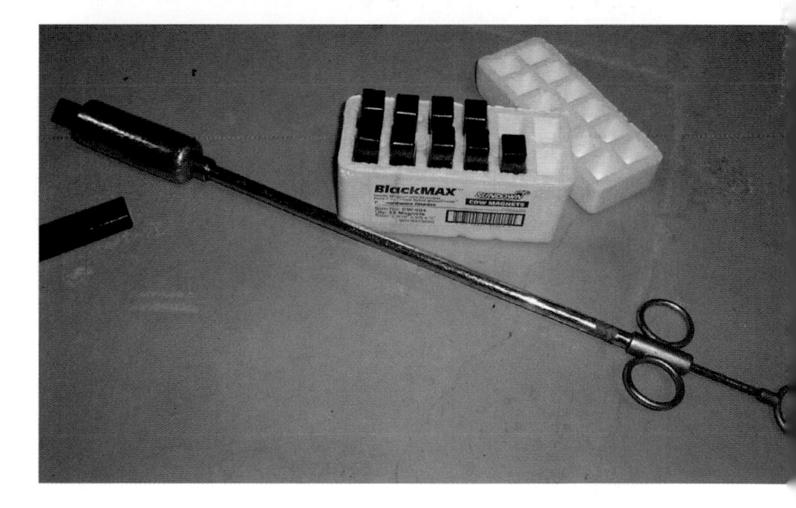

Cow magnets can be administered with the aid of a balling gun. Magnets are strong enough to pick up wire or metal that have been eaten by the cow. The metal sticks to the magnet and reduces the opportunity for hardware to puncture the stomach lining or pierce the heart. The magnet rolls around the rumen for the cow's entire life and does not pass through the rest of the digestive system.

approach any problem. External parasites, including lice and horn flies, can be controlled with powders and other products readily available.

BRUCELLOSIS

Brucellosis is a bacterial infection that causes abortion in pregnant animals, mainly cattle and the American buffalo. It primarily affects the females but males can also contract this infection. The infected cows exhibit symptoms that may include abortion during the last trimester of pregnancy, retained afterbirth, and weak calves at birth. Infected cows usually abort only once and develop a level of immunity to further infections.

Brucellosis is a reportable disease and a single suspected case will cause the herd to be immediately quarantined. This may seem extreme but because brucellosis is highly contagious for other animals it will spread quickly, although not all cows carrying the infection will exhibit symptoms.

Brucellosis is sometimes called bangs disease and can be responsible for a very rapid depopulation of animals. Significant financial losses can occur if your herd is affected. You should avoid purchasing animals that do not have visible signs of being vaccinated, such as state-issued, colored ear tags.

The best way to avoid a herd infection is to have your calves vaccinated at the appropriate age and to buy animals that come from herds that have maintained a routine vaccination program. A discussion with your veterinarian about ways to avoid bringing animals into your herd that are at risk will pay dividends later by staying clear of infections.

Whether on pasture or in a confinement feeding program, cattle must have constant access to clean, fresh water. Tanks that can be heated in the winter will keep the water from freezing.

PINKEYE

Pinkeye is the common name for the reddening of the animal's eye caused by bacteria. It is highly contagious and if left untreated it can affect the eyes and possibly render the animal blind in the eye. Typically this condition occurs during the warm months and appears to be transmitted from one animal to another by flies that congregate around their faces. Pinkeye can be treated with powders, ointments, applications of antibiotics, or with the use of homeopathic remedies. If you treat an animal with pinkeye, be sure to use good hygiene during and after treatment to avoid contaminating yourself. Humans can get pinkeye from animals.

RINGWORM

Ringworm is a contagious infection of the skin and is easily recognized by the circular white encrusted spots on the skin. The infection primarily affects young animals and is caused by a fungus, which is more difficult to treat and takes longer to eliminate from the animal than other infections. One of the oddities of this fungus is that it may lie dormant for several years and seem to be gone until conditions become right and you experience another outbreak.

The fungus that causes ringworm can be killed by disinfectants, such as iodine and glycerin that penetrate the scabs and saturate the spots. Outbreaks of ringworm can be prevented by not bringing infected animals into your herd. The infection can be largely controlled by maintaining clean surroundings and well-ventilated barns where animals are kept.

Because ringworm is highly contagious and easily transferred from animals to humans, precautions must be taken to ensure that you or members of your family do not come in contact with infected areas. Using disposable latex gloves when treating the infection and cleaning out the pen where infected animals live will help prevent spreading the infection.

WARTS

Warts are caused by four different types of the papillomavirus and usually affect young calves and heifers under two years of age. They are usually more of an appearance problem than a physical problem. Their sudden appearance usually coincides with a sudden disappearance several months later. Vaccines are available, but most warts will disappear on their own if they are left alone.

Cattle oilers containing products to discourage flies and insects can be placed in gateways or alleys where cattle pass. Rubbing their backs with these liquid products will give comfort and relief to cattle.

POSTSCRIPT FOR HEALTH PROBLEMS

Discussing these problems is not meant to frighten you from pursuing a program of raising cattle but is meant to alert you to their existence. These are only a few of the health issues that can have possible impacts on your herd. Generally speaking, these problems do not exist on all farms and may not be a cause of great concern.

Pastured cattle appear to require less veterinary assistance and exhibit fewer illnesses because they live in the open air unencumbered by effects of unnatural housing, which allows their healthy bodies to ward off infections. The good manager learns as much as he or she can about potential problems and has an understanding of how to react when they do arise. This is good husbandry.

CONSIDERATIONS

If vaccines, antibiotics, or some other products are to be injected into your animals, consideration must be given to the location of the injection site. All injections should be given subcutaneously (under the skin) when possible because injection needles leave marks and lesions if pushed into the muscles of the animal. Muscles in the neck can be used if it is necessary to give intramuscular injections. Injections into the hind quarters—the rear legs, rump, or hip—should not be made because of the marks left in the muscle when it is sold for market or slaughter. You should always read the labels of any product used on your animals and follow the directions indicating the amount of time needed to withhold the animal from market or slaughter.

It is important to keep accurate treatment records for your animals because it is difficult to remember the exact details of a single treatment six weeks later. Good records also help identify recurring problems that may need to be addressed by changes in management, treatment, or the removal of an animal with a chronic problem from the herd.

RECORDKEEPING

Paperwork is not necessarily a favorite job of many, but keeping good records of your cattle business will help you make intelligent decisions and identify potential problems. There are several computer programs that simplify herd recordkeeping, or perhaps written records may be adequate for your needs. Most people do not have a perfect memory and the best way to remember something is to write it down. As the size of your herd increases, so does the need for accurate records.

The range of records you keep may include breeding records, calving records, vaccination records, treatment records, animal identification records, and anything else that you consider useful. You can use names, ear tag numbers, tattoos, freeze brands, letters, photos, or any other common approach to identify an animal in your records. Whatever makes sense and seems logical to you is most important because these are your records to follow. It is useful, however, if the rest of your family can follow your system, so keep it simple and updated.

If you have purchased animals to begin your program, the best time to place some form of ear tag identification on the animals is either when they have been loaded at the point of purchase or as they arrive at your farm. The few minutes that it takes to put tags into their ears will make their identification easier later on.

Generally, one tag placed in either ear is sufficient for each animal. It is easiest to place an ear tag on a newborn calf if you can get close enough to it soon after birth. Sometimes this is not feasible so you can wait and tag the calves when you run them through the chute for vaccinations or castrations. Even though vaccinations are done several months later, the calves will stay close to their mothers during this time and it will be easy to identify which calf belongs to which cow.

Although ear tags are commonly used, they get lost, the number fades, or the tag become difficult to read because of mud, manure, or age. Freeze brands can be easily used for identification and last the lifetime of the animal. Freeze branding alters the hair color and subcutaneous follicles due to the extreme cold of the iron. If properly applied, this form of branding freezes the follicles so they turn white. The animal needs to be restrained in order to do a proper job.

Using photographs in making individual charts has the advantage of placing the animals at one point of time. As the animal grows and matures, an early picture will provide a benchmark. Photographs may work as a form of temporary identification until you have time to run the animals through a chute and apply ear tags.

A record form that can be used for the lifetime of a calf is shown in this chapter. You can start with the mother and fill in the information that is known to that point. The outline of the animal allows you to sketch the color markings of the cow or calf. You may substitute a snapshot for the sketch and have a permanent visual record of that animal. This form can also serve as a breeding record to keep track of the dates the cow or heifer is bred, either by artificial insemination or by natural service with a bull. This record more closely pinpoints the date for her calving and will help you watch for signs of calving nine months later.

If artificial insemination is used, the breeding date will be more accurate than if a bull is used. When using a bull, particularly if you can't observe the breeding moment, writing down the beginning date the bull has access to the cows and heifers will give you, at minimum, a starting date for the arrival of the first calves. However, this leaves the ending calving dates open because some cows may have been bred several times by the bull and may not fall within the calving period of most of the other cows in the herd. Keeping and maintaining accurate breeding and calving records is an inexpensive management tool that can help you make long-range decisions.

As each calf grows, its individual record stays with it and after several years you will have developed an extensive reproductive file on your animal. You will be able to track cows and their calves, know how many of each gender they produce, and determine which cows appear to produce better offspring. These records may factor into your considerations about retaining some cows and their offspring for future expansion because of desirable traits they exhibit. Other cows may not produce quality offspring you wish to keep. Your records can identify those you can afford to cull from the herd.

Vaccination records are helpful if there is a question about the health of an animal. Vaccinations can help maintain the health of your herd, and the vaccination record can help eliminate possible problems if an outbreak of a disease occurs in your area.

Most vaccines need to be administered by a licensed veterinarian and the animal must be identified in some way on the official report form and your identification records will help. The owner receives a copy of this form that should be filed in your vaccination record system. It is important to retain any of these forms for work done on your animals. A separate book or the use of a form where you can make a notation in a remarks column about vaccinations can serve the same purpose.

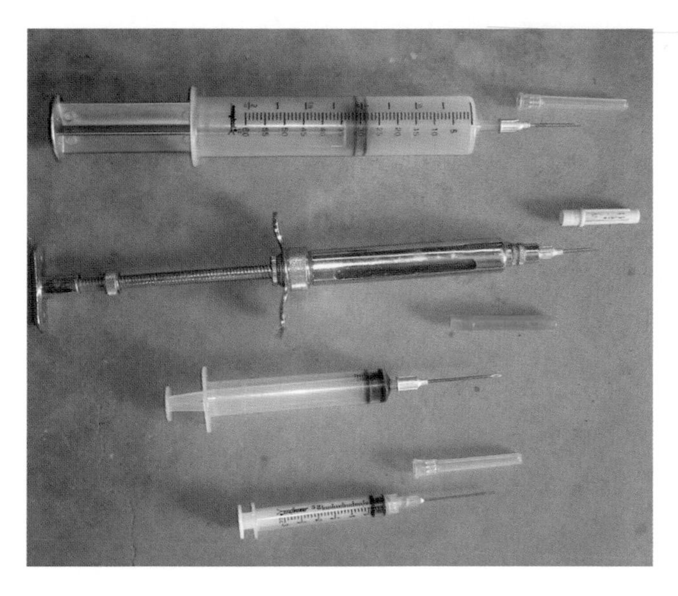

Injection syringes come in several sizes depending on the specific need and dosage. Each needle should have a protective cap. Needles should be discarded after each use to prevent the spread of body fluids from one animal to another.

Keeping a record of treatments of your animals is a good management tool that can help with culling decisions. Identifying which animals have been treated and the reason will help you watch for recurrences and provide an opportunity for prevention. This type of record may be as simple as developing a single sheet of paper for each animal and writing the treatments as they occur.

Your records will most often be used in your management program, but there may be instances when they are required by government agencies if outbreaks of highly contagious diseases occur. Having detailed records available may help with subsequent investigations. If this is not a significant issue, you will have peace of mind knowing that your records are as accurate and complete as possible.

You will become a good manager and develop good husbandry skills with practice and an attention to details. Subscribing to breed publications and general farm newspapers will provide you with the latest information on many subjects relating to your farm and cattle-raising operation.

CHAPTER 14

MARKETING THE MEAT— SAYING GOODBYE

The end goal of any cattle-raising operation should be marketing your animals. Without a plan to selectively remove excess cattle, your farm will become overstocked by increasing animal numbers. After animals arrive on your farm, either through purchase or birth, there are three routes by which they will leave: selling them for slaughter, either for yourself or to a packing house; selling them to someone else; or death.

It is understandable that marketing your animals may be a time of hesitation by you and your family. Having spent considerable time working with your cows and calves, you may develop an attachment to them and regret having to take the next step. Can you say goodbye to your animals? The way you approach your business at the beginning can help turn these ambivalent feelings into a satisfying and positive experience.

Animals have always made a significant contribution to the welfare of human societies by providing food, shelter, fuel, fertilizer, and labor power. One of the many

Cattle are unique animals because their rumen can transform plants unsuitable for human consumption into products humans can utilize, such as meat and milk.

Cattle develop normal behavior traits when raised with other animals. They are easier to handle and are not as likely to pose a danger to others in the herd.

unique aspects about cattle is that they are a renewable resource because they reproduce themselves for another generation. During their lives they utilize other renewable resources, such as grass.

Cattle can parallel human existence in several ways. They serve as partners in reproducing themselves and as companions or social beings. Research has shown that animals raised in isolation do not develop the normal animal social skills as those raised with a group. Animals raised in isolation can be a danger to other animals or their handlers.

Claims have been made that raising cattle is a waste of land and resources that could otherwise be used for producing grain crops that could feed hungry people in a more timely manner. While it is essentially true that it is more efficient for humans to eat plants directly than to allow animals to convert it into human food, this only applies to those plants and plant products that humans can utilize.

Over two-thirds of the forages fed to cattle consist of substances unusable for human consumption. The ability of animals to convert these feedstuffs into food products humans can use greatly improves the diet and quantity of food available for different societies. Whatever inefficiencies there may be in the conversion of grass into human food, it still represents the best way of using those plants that might otherwise be wasted. The claim of wasting land and resources does not take into consideration the costs of cultivating row crops in terms of energy used to till the soil or the erosion caused by ripping up grassland and growing crops that do not have a soil-retaining capacity.

There is also the ethical question of what would happen to the cattle if we did not raise them? Would they

Soil erosion must be controlled when row-crop farming. It is estimated that 2 million acres of topsoil are lost every year in the United States due to ineffective erosion controls, salinization, and waterlogging.

become extinct? If that is the end result of discontinuing raising cattle, isn't it fair to say that we would be contributing to the elimination of another species from the face of the earth, particularly through deliberate intention? Cattle exist because we choose to raise them.

These moral questions of raising cattle can only be answered by those who actually work with them. Outside opinions and claims can heighten the discussion but, in the end, it is those who choose to work with animals that must be left to decide what is to be done. However, once you undertake the challenge of raising cattle, you are under an ethical obligation to provide for them as best you can, treat them in the most humane way possible, and provide a respectful and thankful end of life experience at the time they are marketed or slaughtered.

Today we have choices that our ancestors did not. We have machines to till the soil where our ancestor's relied on animals to provide power. We have food products available that are produced in many other countries and are available year round in ways that our forebears didn't. We have replaced much of the fertilizers from animal manure with synthetic products based on oil.

There are plenty of tradeoffs in whatever you do. In many ways the lifestyle we experience today is oil-based and not a renewable resource. Grass is one of nature's most abundant products and as a renewable resource it costs less to have cattle harvest the grass. Somewhere in between lay the answers that we can live with. Thinking through your objectives will help define what those answers might be for you and your family.

One of the defining ethics for cattle owners is to be thankful for the lives of their animals and the purpose for which they were placed on their farm. This is an authentic discipline for the stewardship of any animal placed in your care. If animals can experience fear and fright it would seem that they can also experience affection and love from their caretakers, which is something akin to human emotions. Although it is not believed they experience it in the same way or to the same extent, it is our obligation to be thankful for their presence in our lives. Otherwise they become nothing more than disposable beings.

Responsible animal welfare is necessary for the long-term sustainability of the livestock industry. Consumers want assurance that animals are raised humanely. Humane treatment shows pride in the care given your animals.

The way animals are handled has a direct impact on the quality and quantity of meat or dairy products they provide. Responsible care makes good economic sense as well as reduces the levels of injury, bruising, and stress. The healthier the animal, the better the quality of the product, and the more profitable they can be.

MARKETING OPTIONS

If your objective is to market the animals that have reached market weight, you have several options. Different marketing techniques have been used for many years including annual feeder cattle sales, local auction markets, and private treaty transactions between producers and order buyers. Auctions have the advantage of establishing livestock prices through competitive bidding, although over the past decade these have been in decline as electronic marketing such as video auctions and the use of Internet markets have been introduced.

Cattle exist because we choose to raise them. Because they are good at utilizing areas where mechanical harvesting is extremely difficult or impossible, cattle are an asset to humans by using areas that would otherwise go to waste. Cattle also replenish grasses with their manures.

PLAN FOR YOUR MARKET

Choosing the time and method of marketing your animals can increase your profits. Too often cattle growers sell their animals at the most convenient time rather than the most profitable time. To become a good cattle marketer, you will benefit from understanding the marketing system and how your cattle prices are determined and then review the options available and decide which to pursue. Bear in mind that it is the consumer that eventually has the most impact on market prices because of their tastes, attitudes, and sense of their family's welfare.

VALUE-ADDED PRODUCTS

Research has shown that the average cow-calf operation in the United States has about 50 mother cows. Your beginning beef program may or may not fit this description.

There are options if you are unable to sell a large number of animals each year. A shift has been occurring from commodity production (where small producers sell through auction markets and feedlot buyers) to direct selling a value-added product. This option may work for you because by offering a value-added product, you can bypass traditional routes and capture more of the profits for yourself.

Selling cattle as a value-added product for, say, $25 per 100 pounds over market price is more profit for you. The best way to achieve a better price is to sell in a way that consumers feel they are getting a better meat product. If consumers feel they are getting a better product then they are probably willing to pay more than if they buy a lower-quality, lower-cost product of the same kind.

Value-added selling may require some marketing skills on your part and finding a niche market where you can sell your meat products. Finding your own customers generally requires more effort than traditional marketing systems but the rewards can be greater.

Some salesmanship may be required in explaining the advantages of your beef to potential customers. Why is it better? It may be because you raise them using a grass-based program or they are pasture-grazed to enhance the eating quality of the meat. You may use other concepts that capture the attention and imagination of potential customers, such as certified organic. Value is created when a product meets or exceeds the customer's expectations.

Niche markets for such things as natural, organic, green, or pasture-grazed are becoming recognized by customers for the healthy conditions under which the animals are grown. Marketing groups may be available to sell your product. Most county agricultural extension offices can provide you with contact information with groups in your area.

Value-added products can provide more income from your cattle-raising enterprise. Customers will appreciate your food products if they have a consistent quality.

Developing a label that identifies your product in the market or to your customers is a simple way of reinforcing your name and making customers aware of the program under which you raise your cattle.

BEST TIME TO SELL

The best time to sell your cattle is when they reach the target weight you have determined for your production system. There may be other factors to consider in timing the sale of your animals, such as the time of year, the market cycle, and the uniformity of the animals to be sold.

If you are selling feeder calves, it is generally more profitable to feed these calves through winter and market them in early spring rather than selling them in the fall when they are lighter in weight and smaller in frame size. However, the increase in price received will have to be weighed against the cost of raising them for that extended period of time.

The market cycle can have a huge impact upon your profits. Try to avoid selling at a time when many animals are expected to reach market at the same time. It may be worth waiting several weeks to sell your animals if the market prices start to rise, as this may indicate a leveling off of animals entering the packing houses. Familiarizing yourself with published cattle statistics and market trends can help you identify the best potential market times during the year.

MOVEMENT OF YOUR ANIMALS TO MARKET

Moving your animals to market in the easiest and fastest way possible will pay huge dividends with the quality of meat you receive. Handling them calmly and avoiding stressful situations when loading will keep cattle from releasing large amounts of adrenalin into their systems. It has been found that a large release of adrenalin into the bloodstream of an excited animal alters the texture of the muscles. During a frightening event, the alteration and contraction of the muscle fibers causes the meat to change its consistency of texture and taste. Adrenalin eventually clears of the animal's system but generally not in the time between loading, hauling, and slaughter.

Developing a corral for loading cattle that moves them in a relaxed pattern will help defuse the fright reflex. The use of shockers, clubs, shouting, and other excitable methods when loading animals should be avoided at all times.

You can haul your animals to the harvest site if you have a cattle trailer or you may need to hire someone with a trailer to transport the animals for you. If you are hiring transportation, make certain the hauler understands your concerns about animal movement and treatment once the truck arrives at the harvest facility. All of your work can be negated if the animals are mishandled or mistreated while being unloaded in your absence.

MARKET YOUR OWN LABEL

Capturing a niche market can have financial and personal rewards, especially when you develop your own label for your meat products. There are labeling requirements but they are not so excessive to prevent you from entering this market. Private labels are becoming more prevalent as producers with small numbers of animals find it to their advantage to use such production practices as grazing programs as a selling point for their products. Since you may raise fewer animals, you can provide more individual care and turn this into a solid marketing tool to identify your farming practices.

You will need access to a harvest facility that can guarantee that your animal's meat will be separated from other meat processed the same day. The workers can explain the costs involved in processing and packaging your meat, which can be stamped with your label for proper identification. Be sure that the facility is state- or federal-inspected.

After the meat has been processed, you will need adequate frozen storage facilities on your premises. Factors that influence the amount of storage space needed for the meat include the total pounds of meat from each animal and the number of animals processed at one time.

Developing your label and storage facilities should be done well in advance of harvesting any animals. Your calculations should be made prior to purchasing storage equipment so it is in place when your meat arrives. When your label and storage system is complete, it will be in place to use whenever you have an animal harvested and processed for sale.

If you plan to sell and deliver meat to customers who live some distance from your farm, you will need a storage container that can maintain frozen temperatures while transporting the meat. Contact your state department of agriculture for details on how to start direct-marketing meat. They can advise you on all regulations and requirements involved in labeling your products.

ADVERTISING YOUR PRODUCT

Advertise your product when you have meat to sell. How you approach this, whether by word of mouth, print ads, or some other means, will determine how long your meat stays in storage. If you have more animals growing on pasture, it would be to your advantage to move the meat quickly and have room available for the next round.

Help is available with your marketing and advertising program if you don't feel that you can handle it alone. Be sure to highlight the emphasis you place on a certain production system; it is appreciated by consumers.

SAYING GOODBYE

The final disposition of your animals may cause some ambivalent feelings for you and your family. Raising animals

Storing meat at home requires a freezer with capacity for all of the processed animal products. Maintaining appropriate temperatures will prevent spoilage and allow you to store the products for long periods of time.

that you work with on a daily basis can be an emotionally rewarding experience for everyone. There is a certain type of communication that can exist between animals and humans and to be part of that experience is an exceptional encounter with a world on a completely different level.

Like humans, these animals develop their own personalities and by extension, project their mannerisms to those who work with them. There may be cows that instinctively know what you want them to do and there may be cows that walked across the pasture just to say hello and receive attention. It can be difficult to let these animals go and turn them into food. But, as mentioned at the beginning of this chapter, this is their cycle of life and you are part of it. In essence, you are helping them fulfill their destiny and one that was done with concern, caring, love, and a sense of pride in providing an atmosphere where they lived comfortable, healthy, and quality lives. There is no shame in that and you can find great satisfaction in achieving these ends and by honoring their presence.

A chest freezer can be used in some instances for transport of meat. Check your state regulations for further information.

Appreciating your animal's contributions to your livelihood is a moral obligation for anyone raising cattle. Working with cattle on a daily basis can be an emotionally rewarding experience. To honor their presence in our lives is to be thankful for all they have given us.

CHAPTER 15

MILK— NATURE'S MOST PERFECT FOOD

Milk is a nutritious food that can be turned into 30 primary products. Whether you milk 3, 30, or 300 cows, you can take advantage of your cows' ability to turn grass into milk to make a living.

Cow's milk has been a food source for humans since prehistoric times. Its nutritional value is extremely high and it can supply more of the daily nutritional requirements of the human body than any other single food produced on the planet. Milk is a versatile substance with more than 30 primary products made from it and several subproducts derived from each of those groups. No other form of agriculture approaches the number of products derived from one primary source.

Producing milk can be a satisfying and rewarding occupation because it blends work with animals and crops as you assist in the transformation of grasses and other feedstuffs into an edible food product. Whether you own one cow or several hundred there are three basic options for the milk you produce: drink it, sell it, or turn it into another product. The more cows you milk, the more pounds of milk you will have to market and present new possibilities for your farm.

Milk production involves a complex process, but in the simplest terms, grass or other feedstuffs are converted into milk starting with a fermentation process inside the cow's stomach. This intricate conversion of fiber to liquid by way of four stomachs is one of nature's most fascinating processes. It is identical to that of beef cows, which also convert grass or hay into milk for their calves on a small scale. The physical nature of a beef cow transforms the same feedstuffs in the rumen into muscle rather than large quantities of milk.

The milk production process starts when a cow chews and swallows grass as it is grazing or eating hay or other feedstuffs. The rumen of a beef or dairy cow does the same job of breaking down the large plant particles into smaller ones by grinding and chewing. The chewed food, or cud, is eventually passed along to the reticulum, or second stomach, where the chemical processes start breaking down the fibrous materials to be converted into amino acids, proteins, and other components that are secreted to the udder, where they undergo the final transformation into milk. This

Depending on the circumstances in your area, you may decide to sell your milk to a processing company or a cheese factory. If you have the interest, you can process milk on your farm and sell it.

abbreviated explanation can be studied more completely with the aid of many books available on the subject.

MILK-PRODUCING COWS

A dairy farm typically restricts itself to milk produced by dairy cows such as Holsteins, Guernseys, Ayrshires, Brown Swiss, Jerseys, Milking Shorthorns, and others. Milk from beef cows is generally used to feed their calves.

A dairy cow produces the largest daily milk volume at the beginning of her lactation. This milk-producing ability, or lactation, is initiated when the cow has a calf. The initial milk is called colostrum and contains immunoglobulins that help the newborn calf develop resistance to diseases.

You can develop a small dairy herd of one or two cows that can provide all of the milk your family can drink in one day. Today a typical cow in the United States produces 50 pounds of milk each day or over 16,000 pounds each year. Multiplying that volume by the number of cows you have will give you some idea of the volume you can produce from a small herd.

The number of specialty and artisan cheese plants has increased as more farmers find cheese-making to be a rewarding career choice. On-farm cheese factories are one option to produce a value-added milk product.

REGULATIONS FOR PRODUCING MILK

Certain equipment and facilities are needed for milking dairy cows whether your herd will be small or large. Small herds do not require the extensive facilities needed for larger herds, nor do they necessarily require the vast amount of oversight by regulators. If you produce milk only for your family, the regulations are few. However, if you sell the milk, other requirements and regulations take effect.

By determining the size of herd you want, you will be able to anticipate some of the requirements and regulations that will affect your dairy enterprise. Information about rules, regulations, and guidelines for producing milk for public consumption is available from your state department of agriculture.

MILK PRODUCTION OPTIONS

Depending on your goals, interests, and financial situation, you may decide to produce milk for your family, for sale to a milk-processing plant, or for the production of cheese. There are other options for your milk but those involve greater financial resources and more specialized equipment. If you are in the position to invest a large sum of money into a dairy enterprise, be sure you have capable advisors and managers involved.

Selling whole milk may be the easiest option for your business, especially if you are just getting started. A license issued by your state's department of agriculture is required to sell milk from your farm to a milk company, processing plant, or cheese factory.

There are regulations to meet and a state milk inspection must take place prior to selling milk off your

Smaller equipment is needed with a farmstead cheese factory because of lower amounts of milk available. Specialized training is required in some states before being licensed as a cheesemaker.

Artisan cheeses include a vast array of names, styles, and tastes. An individual cheese maker can produce whatever type of cheese he or she feels comfortable making and marketing.

farm. This inspection includes reviewing major items such as areas where the cows are housed, the milking equipment and milk holding tank to be used, the water supply and well, the sanitation program, and barn cleanliness. Arrangements with a milk processor, a cheese factory, a milk cooperative of which you become a member, or whomever will be buying your milk need to be made before it leaves your farm.

Consumer interest in buying raw milk directly off the farm is growing. Raw milk has not undergone pasteurization. Some people believe the pasteurization process destroys many of the enzymes that occur naturally in the milk and they believe that drinking unpasteurized milk is healthier. At the time of this writing, raw milk sales are legal in 28 states across the country. In some of the remaining states, raw milk is available through a cow-share program where shares in a cow are sold to customers who have a legitimate ownership of the animal and can receive milk from her such as you can for your family.

Historically, health concerns arose over drinking raw milk due to the prevalence of brucellosis in dairy cows. The almost total eradication of this bovine virus has greatly reduced the risk of drinking unpasteurized milk. However, other health concerns such as E. coli and salmonella contamination require that strict sanitation and herd health standards be maintained. With healthy animals and good sanitation, you may find the option of selling whole milk to be part of your business plan.

More dairies are selling milk from their farms as retail businesses. This takes an investment in time, equipment,

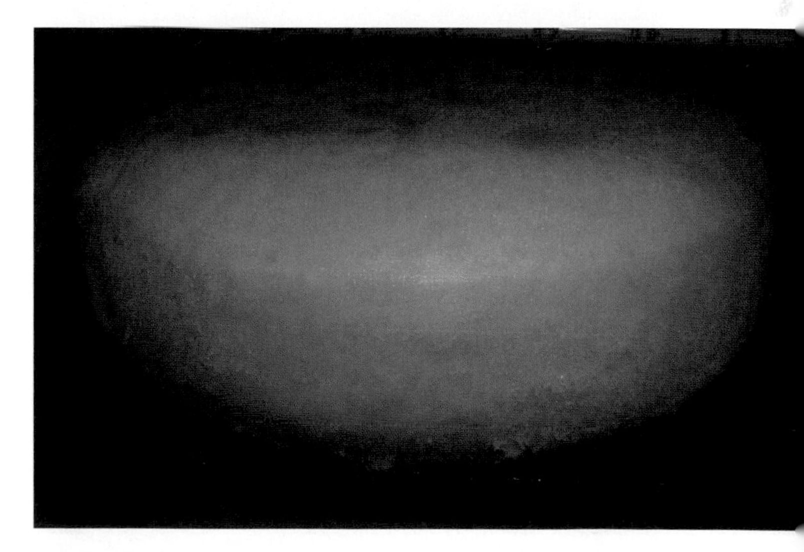

The end result of an individual farm cheese-making business will be a unique product that can be marketed to restaurants, stores, or direct mail customers.

and management. If you are close to an urban area, marketing milk under your own farm label may be an option.

You can sell your milk to a cheese factory that specializes in different cheese products. There are small factories that specialize in organic cheese, rBGH-free cheese, or other artisan methods for making unique cheese varieties. These are often value-added products that can provide a greater return on your milk than the more conventional route of selling milk as a commodity.

Cheese making is an option that is gaining popularity on some small farms. Farmstead cheese making requires a separate facility and an investment in specialized equipment and training. If you have any interest in making cheese from your milk, this may provide you with a more profitable option than simply selling milk.

A boiler system is needed to create and maintain high water temperatures for cooking the milk and for cleaning purposes. Efficient systems provide hot water quickly when needed, with an ability to cool rapidly as well.

Small farms do not need to produce a lot of cheese to provide added income. On average it takes 10 pounds of milk to produce 1 pound of cheese. You can easily calculate your potential cheese production by knowing the volume of milk you want to produce.

A cheese-making facility can be built on your farm and the milk can be pumped directly from the bulk tank to the factory through special tubes. This eliminates the need for handling the milk more than necessary, which may affect its quality.

Farms can incorporate pasture-grazing programs with cheese making by using the milk produced into a value-added product that can boost income. Good planning is essential to construct a proper facility.

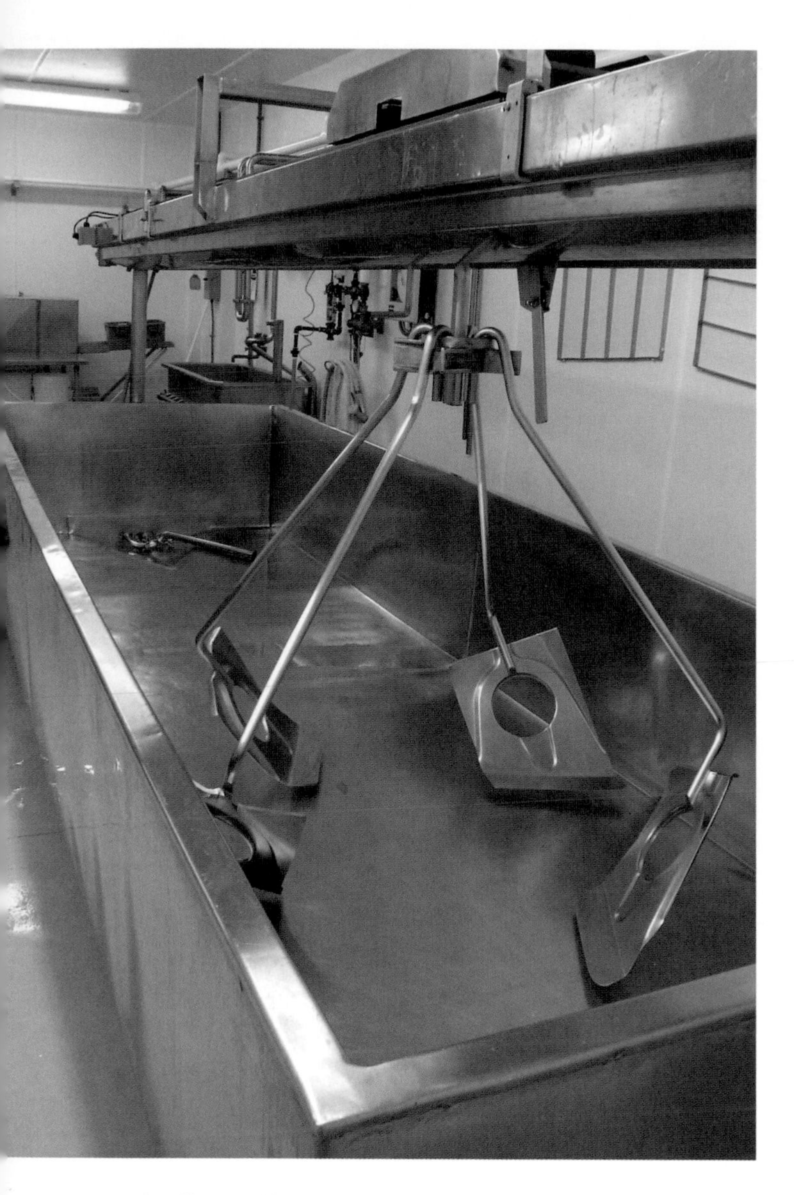

A milk vat with a 9,000-pound capacity will hold enough milk to produce about 900 pounds of cheese. Paddles stir the warm milk while starters and rennet are added to begin the cheese-making process.

Each of these individual hoops holds 10 pounds of cheese. Each stack has a press for squeezing whey from the cheese. Small specialized equipment makes cheese making on a farm possible.

GRASS-FED MILK PRODUCTION

The most economical source of feed for your dairy cows can be the grass growing in your pastures. If managed correctly, pastures can provide the bulk of roughages to feed dairy cows throughout the year. In many northern areas of the United States, grass is difficult to eradicate. It wants to grow in almost any open spot and this persistence is to your advantage.

The biggest advantage of producing milk from a grass-based dairy system is the potential to generate more income from lower expenses than the conventional approach. It is one of the best ways to get a start in dairying. Traditionally, milk has been produced from a conventional system where cows are milked year-round to achieve a steady, even flow of milk. In this system, cows are fed hay, silage, and corn. This is generally a response to a cash flow situation on the farms where a minimum monthly production is needed to help meet expenses.

This cycle has changed over the past decade as more dairy farmers with smaller herds desire time off for themselves and their families by milking only 10 or 11 months of the year and having their cows stand dry, or not milking, for the other month(s). A grass-based system of dairy farming can incorporate seasonal production in ways conventional dairying cannot. Seasonal systems can match the reproductive cycles of the cows and the forage availability of the farm.

This approach can be modified depending on the region of the country in which you live. The goal is to avoid the period of highest-cost milk production while using the least-expensive feed source available. In the midwestern, northeastern, and northwestern parts of the United States, the reproductive cycle of the herd can be timed to match the period of greatest forage availability, which is usually in the spring. In warmer climates, it may be advisable to dry off the cows during the warm, humid summer months and take advantage of the cooler fall, winter, and spring months as the optimum production periods.

Sinks and automatic washers are used to wash and sanitize equipment before and after use. Cleanliness and proper sanitation are essential to producing quality cheese.

CHAPTER 16

DAIRY PRODUCTION—BASIC CONSIDERATIONS AND FACILITIES

In the past, family-owned farms dotted the landscape of major dairy states. Today, the trends toward mechanization and modernization have moved dairy farming into a very capital- and labor-intensive business. Specialized facilities and cattle management expertise are paramount in this type of cattle enterprise. Large herds and multi-million-dollar facilities are becoming more the norm in many dairy regions. However, there are still opportunities to begin a smaller-scale farm, particularly if a value-added market for the milk is pursued.

A farm background is not necessary to raise dairy cattle, but any experience you can gain before starting your own enterprise will be helpful. There are several ways to enter the dairy cattle business:

Enroll in a formal training program at a technical school or university or through your local extension office for beginning farmers to gain knowledge and expertise before buying cattle.

A farm background is not necessary to work with dairy cattle. Starting with a small herd is the best way to begin dairy farming because of a lower initial investment.

Large dairies are considered to be more than 200 cows. A 400-cow herd, such as this farm, requires a large set of buildings to house cows, calves, and milking facilities. Large dairies, while capable of producing large volumes of milk, also require a huge investment in capital, equipment, and personnel.

Work for an existing dairy farm to learn about managing dairy cattle and the business as a whole before starting your own dairy farm.

Assemble your own herd of cattle and house them on someone else's farm until you are ready to launch your own business. Owning cows already in production provides a quick income once you move into your own facilities.

Buy cows, purchase feed, and rent facilities and equipment to get started. Fixed assets, such as land or equipment, can drain your financial position when starting out. Keeping the amount of owned equipment to a minimum allows you to invest in cows that are productive assets rather than machinery that, although needed at specific times, generally sits idle during much of the year.

Buy calves and heifers (instead of adult cows) and start raising them while you are building or remodeling your own farm facilities and installing milking equipment. During this time, also make arrangements to handle the fieldwork involved to grow feed. Decide whether to purchase

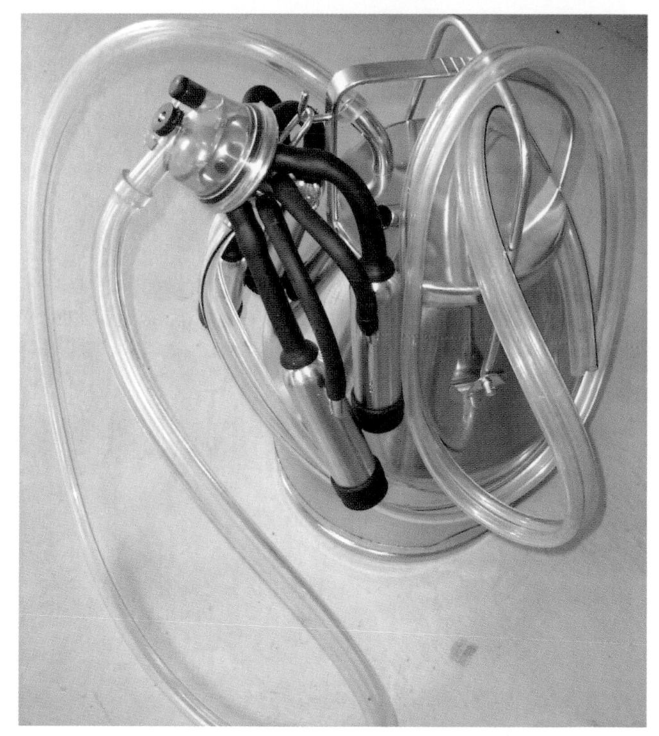

Milk buckets can be used efficiently for small herds. Each bucket is a self-contained milking unit that extracts the milk from the udder. After the milk is completely emptied from the udder, it can be poured into another container and taken to the bulk tank for cooling. The teat cups are dipped into a sanitizing solution before being attached to the next cow.

A flat milking parlor allows the cows to walk into the milking area to be restrained for milking. The raised platform makes it easier for the person milking as he or she can sit to assist with applying and removing the milking unit. Once the milking is finished, the cow leaves the stall and walks back to the feeding area.

equipment, arrange for custom hire, contract to purchase feed, or any combination of these to handle planting and harvesting to feed your cattle.

Your ultimate success in dairying will depend on your experience and aptitude for dairy cattle management and access you have to the capital needed to begin and continue in business. Many farmers choose to begin with a grass-based dairy. Since feed costs generally account for 50 percent of total operating expenses, any way to economize in this area will free up funds to use for other expenses.

Dairying can provide a greater income per animal than any other farm enterprise, with the average cow contributing $2,000 to $2,500 per year in gross sales. However, to establish this type of business requires detailed planning with a well-thought-out plan, no matter what the scale of your farm. Bankers and accountants specializing in agriculture can assist you with formulating budgets and projected cash flow, and can give you advice on other financial considerations. County agricultural extension agents can provide information on accessing services and answer general production management questions.

FUTURE OUTLOOK FOR DAIRYING

Milk supplies 73 percent of the total amount of calcium consumed in the United States. With the increasing recognition of the importance of calcium to the diet of many women and children, it will be a long time before this source of calcium is supplanted by something else.

In areas such as the Midwest and Northeast, there may be more markets for selling milk than selling other livestock. Dairying provides a monthly income, which is different from raising beef cattle where income is seasonal and only upon sale of the animals. As with other farming enterprises, hard work, good planning, and an eye for detail

Individual stalls that cows can use at random are called free stalls. The cows' movement is unrestricted as they walk to eat or lie down. Stalls can be filled with sand, special foam mattresses, or other soft bedding materials.

Cows will seek shady areas during hot weather and will bunch themselves together. Providing shade and access to water is important in maintaining production levels during all types of weather.

Calves that are raised in pens should be housed in well-lit, dry areas free from drafts. Wood shavings make an excellent bedding material that keeps moisture away from where they lie down. Calves should not be crowded into pens that are too small for their numbers.

will help ensure success. Hundreds of cows are not required to be successful. However, if you aren't profitable with 40 cows, chances are you won't be profitable with a larger herd.

MILK HANDLING

Milk has a short shelf life unless it is processed into other products. In order to maintain quality, milk needs to be handled with care as it travels from the cow through processing. Federal and state licenses are required to sell milk and many regulations apply to sanitation, good animal care, and a variety of other issues.

When you apply to sell milk to a processing company or cheese plant, you will become involved with the regulating processes dictated by your state. A state milk inspector will inspect your milking facilities and areas where you wash and sanitize equipment. The inspector will also look at your barns where cows are housed, water sources, storage areas for cleaning chemicals, and other things that could affect milk quality. Inspectors do not examine the feed, equipment used outside the barn, or anything unrelated to milk production.

Check with your state milk inspector or milking system service personnel about the regulations for installing milking equipment and regulations that apply. Inspections should not be viewed as an adversarial situation. The milk inspector is obligated to make sure your facilities are in accordance with the present regulations to insure that the milk you produce is fit for human consumption. Inspectors can be a good source of information and are often willing to help advise you on correct procedures for milking if this hasn't been discussed with you by your service representative.

MILKING FACILITIES

Hand milking was replaced by milking machines decades ago. Milking machines have become essential and reliable equipment on dairies. Milking facilities today range from barn milking (bucket-type milkers are used and milk is transferred to a holding area) to pipeline systems (which eliminate the need for milk buckets) to milking parlors.

A well-stocked dairy cabinet includes antibiotic test kits, iodine, rubbing alcohol, and other products typically used in sanitation. Good sanitation is an important key to reducing mastitis infections in a dairy herd.

The choice of milking system may be related to your own physical condition. Bucket and pipeline milking systems require a lot of lifting and bending. Milking parlors are popular because a larger herd can be milked quickly and efficiently. It is also physically easier for the person doing the milking. Milking parlors come in many sizes and shapes, from the very basic swing parlors to elaborate stationary or rotary structures with many features. There is usually an incremental increase in cost as you climb the scale of parlor structures.

Proper milking equipment is extremely important. It is wise to develop a business relationship with a local dairy supply business. Once installed, your milking system will require routine checks to maintain optimum performance, such as maintaining proper vacuum levels in air lines that ultimately affect the health of your cows' udders.

An inspection of your milking system is required before you are allowed to sell milk off your farm. These inspections can be arranged with the help of the field supervisor for the milk plant or cheese company to which you decide to sell your milk.

Grouping heifers according to size rather than age will reduce the dominance of larger animals and the risk of injury. Raising heifers in open areas requires less work and makes it easier to provide feed and water.

Plastic curtains positioned between the holding area for cows waiting to be milked and the milking parlor can provide a wind break in inclement weather. Cows quickly learn to walk through the overlapping heavy plastic strips.

Tie-ups that are installed for smaller animals for vaccinations, castrations, and ear tagging will work best if the animal cannot escape before being tied. High sides will keep a young animal from trying to climb over a railing.

COW HOUSING

There are a variety of options available for housing your cows. The climate where you live will dictate what type of shelter you will need.

Farms in the Midwest have traditionally used barns for housing cows because of the need for shelter during the cold, harsh winters. Barn interiors usually include stanchions, tie-stalls, pens, or a combination of these options. Free stalls have recently become popular as dairy owners have developed more experience handling cows in cold weather. A free stall system involves individual stalls in rows where cows have access to lie down, usually a short distance from the feeding area. Cows have free choice in picking the stall they want. However, because cows are creatures of habit, many will seek out the same stall each time.

Other methods for housing cows include open dry lots, bedding packs in sheds, and open bedding packs outdoors. Each shelter can be effectively used to give cows a clean, dry place to lie down, which is important for maintaining good udder health.

Cattle can withstand very cold temperatures if they have shelter from the wind, access to water, and dry bedding. Similarly, during hot spells cows need access to shade. Some dairies also provide water sprinkling systems to cool animals. As temperatures and humidity increase, milk production decreases. Providing a place where cows can get out of the sun and heat will help minimize the drop in milk production during the summer.

HEIFER FACILITIES

If you plan to raise calves, separate housing facilities are also needed. Typically, different-sized animals require different-sized pens or corrals. It is important to group similarly sized calves and heifers together because of their social attitudes. When housed together, smaller calves and heifers are dominated by larger animals, causing the smaller calves to get less feed and not grow as well. Keeping calves and heifers of a similar size together allows uniform access to feed.

While heifer facilities do not need to be as extensive as the facilities for your cows, young calves must have access to food, water, and shelter, just like cows. It is important to provide a dry area that is free of drafts for newborn calves through six months of age. Pneumonia in

Individual milking buckets can be made of metal or hard plastic. In many parlors, a plastic bucket is used for fresh cows or those treated with antibiotics to keep their milk separate from the milk from the rest of the herd.

calves and heifers is common in facilities with inadequate ventilation or wet bedding.

MILK QUALITY

Milk quality is directly related to the udder health of your cows. Udder health is contingent upon proper cleanliness at milking time and the conditions of stalls and corrals. Cows need dry, clean bedding materials. You want your cows to avoid exposure to manure, mud, or any other substance that can harbor mastitis-causing bacteria.

Mastitis, an infection and inflammation of the cells within the udder tissue, is caused by bacteria. Mastitis causes a reaction from the cow's immune system that quickly elevates the number of white cells to fight this infection. These cells pass through the udder into the milk, which results in a high somatic cell count (SCC). As the SCC number increases, it indicates the presence of a high level of infection somewhere in the herd. This will require immediate attention for several reasons. A cow with an udder infection is susceptible to a major illness, and depending on the severity of the infection and the type of bacteria involved, it may be life-threatening to the animal if left untreated.

The SCC serves as an indicator that something is wrong with the cow management system or that inadequate vacuum levels exist in the milking system, which is causing problems in the cows' udders. Incorrect vacuum levels will cause fluctuations at the teat end during the suctioning of the milk from the udder and leads to irritation of the tissues, which become susceptible to infection agents.

Milk producers often receive price premiums added to the base milk price when the low target levels of somatic cell counts and other quality practices are reached. Maintaining high milk quality can increase monthly income.

ANTIBIOTIC USE

Antibiotic use, treatment regimens, effectiveness, alternatives, and its residual levels in milk should be studied prior to establishing a dairy herd. If antibiotics are needed for treating mastitis or other health problems, read all label directions and strictly observe all withholding times listed.

Every cow is different in how she metabolizes antibiotics. It's not uncommon for some cows to have their milk test positive for the presence of antibiotics well past the time ordinarily required for withholding her milk from the market. If you are treating a cow with antibiotics, it is very important that her milk does not go into the bulk tank until it tests negative for the presence of antibiotics.

A sample of your milk will be taken from the bulk milk tank by the milk hauler prior to loading it onto a

A hip restraint is a humane way to keep young cows from kicking at a milking machine and provides a measure of safety for the person milking the cow. Many first calf heifers find milking uncomfortable and will kick at the machine to remove it. A hip restraint harmlessly prevents them from lifting their leg to kick the machine off their udder.

truck. This sample is tested before the truck is unloaded at the plant or cheese factory for the presence of antibiotics, which adulterates the milk and makes it unusable for human food consumption either as milk, cheese, or other dairy products. The sample is also tested for the SCC to determine the quality of the milk.

Antibiotic test kits are available for on-farm use. Having these on hand and familiarizing yourself with their use will help to avoid major problems. These precautions are important because if any antibiotic residue is found in milk coming from your farm, the entire truckload of milk is rejected at the milk processor. Not only is income lost for your milk, but some plants may require the dairy that supplied the antibiotic-tainted milk to purchase the entire truckload if prior notification of a problem has not been given.

The use of antibiotics on certified organic dairy farms is prohibited. If this is your market, you need to seek alternative treatments, such as homeopathy, which can be useful and beneficial in preventing and alleviating the effects of mastitis.

STRESS FROM FACILITIES

Stress is a contributing factor in udder infections and a cow's general health will largely determine her quality of milk. Removing stress, whether it comes from crowded facilities, trying to produce more milk than a cow is capable of, or confinement without access to outside pastures, can help diminish the incidence of mastitis. It is to your advantage and the health of your cows to maintain a good milking routine with adequate sanitation, clean stables, and the elimination of stress.

PURCHASING COWS

Dairy cows can be purchased in ways similar to beef cows; individually or in groups, privately or publicly. You want to find healthy, mobile, pregnant cows that will be able to produce milk for you for several years. If you do not feel informed enough to buy dairy cows on your own, it may be worth the money to hire a reputable cattle dealer.

A simple way to calculate the cost of a cow relative to expected income is shown here. Although these are estimates, the formula can be used by inserting your current prices for milk and cows. As of the writing of this book, a dairy cow that is considered physically sound, in good health, and pregnant can be valued between $1,600 to $2,500, depending on a variety of market conditions such as cattle availability and milk price. For this example, the value of a cow is $2,000. If she produces 15,000 pounds of milk per year and the milk price received is $12 per hundred-weight (100 pounds) then,

15,000 lbs = 150 hundredweights (cwt.)

150 cwt. x $12.00/cwt. = $1,800 gross income per year for milk

1 heifer calf = $500 or 1 bull calf = $200

If the cow bears a heifer calf, then the value of products the cow produced for that one year would be $2,300 ($1,800 + $500). However, the available gross income will be $1,800 unless you sell the heifer calf.

In a typical situation half of the milk income is needed to cover feed costs during the year. Therefore,

$1,800 x 0.5 = $900 in gross milk income for the year

According to these figures, the example cow needs to stay in your herd 2.3 years to recover her original cost. However, this total will have to be applied against other farm expenses, such as veterinary costs, insurance, utilities, depreciation, and so forth. The total operating expenses of your farm can be broken down into a per unit (per cow) cost so you know how much milk each cow needs to produce to reach a breakeven or positive profit level.

Agricultural financial advisors can help you develop a farm plan to determine the prices you can pay for cows and still be profitable. While economic advice is calculated from the single issue of pounds of milk produced, there are other factors that can affect a cow's profitability. When looking at a cow to purchase, take note of the following areas: feet that may need to be trimmed or treated for soreness; the structure of the pelvis, which can cause ease or difficulty in calving; the strength of the heart girth or front end, which may affect the general health of the cow; and teat structure, which affects udder health. All of these criteria should be part of your deliberations when you purchase a cow. If needed, seek help in determining what constitutes a good sound animal. Breed organization personnel or independent groups, such as Animal Analysis Associates (aAa), can help advise you. Buying sound cattle and putting them into a comfortable facility will help you produce quality milk. In the long term, this is the most profitable way to dairy farm.

Tie-stalls allow more freedom of movement for cows than stanchions. Cows are tied with neck straps, which are attached to a chain at the front. Less restraint of a cow's movements in a stall often results in less injury.

CHAPTER 17

DAIRY CATTLE—CHOOSING A BREED

Brown Swiss are large-framed cows, similar to Holsteins, but have a higher average fat and protein milk content. They perform well in rugged conditions and adapt very well to different environments.

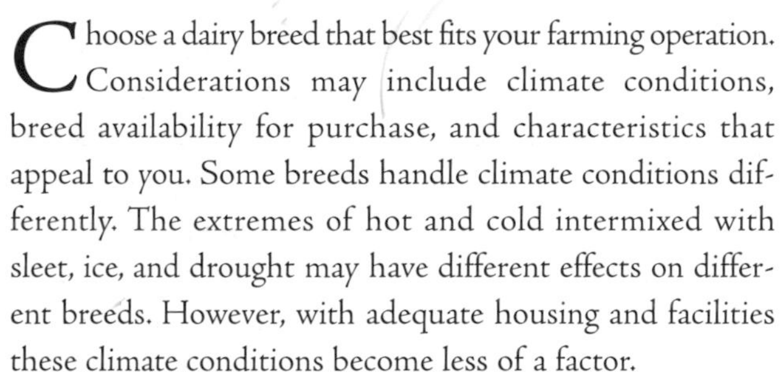

Choose a dairy breed that best fits your farming operation. Considerations may include climate conditions, breed availability for purchase, and characteristics that appeal to you. Some breeds handle climate conditions differently. The extremes of hot and cold intermixed with sleet, ice, and drought may have different effects on different breeds. However, with adequate housing and facilities these climate conditions become less of a factor.

Total acreage and the kinds of pasture and forage crops that are raised on your farm or purchased locally may also influence your decision. Breeds vary in amount of feedstuffs consumed, grazing ability, and adaptability to varying climates. Some breeds have a higher average butterfat and protein content in their milk. Milk prices are typically based on protein and butterfat content, in addition to the volume of milk sold.

Dairy farmers usually seek to raise cows that produce large quantities of high-quality milk, calve yearly, and live long and productive lives. Increased milk production and lower input costs increase profits. Dairy cattle are efficient converters of grass and forages into milk. Therefore, many beginning farmers graze their animals on pasture because it is a highly economical feed source. Some breeds are better grazers due to their physical structure, and their adaptation to the conditions and climates where the cows live.

MILK PRODUCTION

Genetic ability to produce large quantities of milk is important, and management is a key factor to realizing this potential. Giving birth to a calf starts a cow's milk-production cycle. This is called freshening and it initiates what is called a cow's lactation. Once a cow starts her lactation, she produces milk quickly and for a long duration.

Most dairy producers manage cows to calve every 12 to 14 months, which is called reproductive efficiency. Toward the end of a cow's yearly production, the amount of milk produced daily tends to drop. This is partly due to the cow being pregnant. The cow's body signals a time to stop producing milk in the present lactation in anticipation of the calf's arrival. Milk volume is also affected by feed availability and quality, particularly if a cow is on pasture.

The longer a cow lives, referred to as longevity, the more calves she will produce and the more profitable she will be. Until a cow freshens for the first time (around 24 months of age) she will not produce milk. However you will still be spending money to feed and care for the cow during this time. In calculating dairy profitability, the costs of feeding a heifer to freshening must be offset by her number of productive years. Typically it takes four years for a dairy animal to produce enough milk to pay for the costs of raising her the first two years of her life.

A cow that stays in the herd for only one lactation is a losing proposition for its owner. A cow producing for at least two lactations will have had two calves. With a 50/50 heifer calf to bull calf ratio, chances are the cow will have had at least one heifer calf to later replace herself in the herd. Having only one calf reduces the chance of producing a replacement in two years and is the reason reproductive efficiency is so important.

MILK COMPONENTS

Butterfat and protein content are important components of milk. If you live in an area where component pricing is used, the breed you choose may be of prime importance. Some dairy breeds have a higher butterfat and protein content in their milk. For example, you can capture markets for specific cheeses or butter quality by selecting a certain breed of cow to raise. Feed may also influence some of these qualities. However, genetics play a more important role.

Some breeds are larger than others and this may be a factor in the breed you choose to raise depending on your feed situation and available housing. Larger animals not only eat more feed, they also require more housing space. In a barn or free stall, some breeds require a larger stall to accommodate their body size and reduce their chances for injury.

Guernseys have made a resurgence as a breed with the increase in the number of dairy grazing programs. They are excellent at converting forage to milk and have docile dispositions.

DAIRY BREEDS

Traditionally, dairy farmers in the United States have raised five major dairy breeds over the past century: Holstein, Guernsey, Brown Swiss, Jersey, and Ayrshire. Within the past 20 years the Milking Shorthorn breed has transformed itself from a dual-purpose breed used for both meat and milk to one specifically bred for milk production. The other Shorthorn branch has continued as a beef breed. Interest in grazing cows on pastures has been on the rise for over a decade and several other less-prevalent dairy breeds have risen in popularity because they are better grazers.

Generally most farmers breed their dairy cows to bulls of the same breed, although some farms may have several breeds of cows within their herds. Staying within a breed is called purebred breeding, but technically a purebred animal is one that is registered in a breed association and has a verified ancestry.

Crossbred dairy cows are being promoted because of the hybrid vigor that some breeders believe helps produce a better dairy cow. Hybrid vigor is a result of crossing two or more breeds to try to produce an animal that receives the best qualities of each of the parents, such as the high butterfat and protein percentages from the Jersey and the milk volume of the Holstein. Whether you choose one breed, several breeds, or crossbreeding, there are many options to consider.

AYRSHIRE

The Ayrshire breed originated prior to 1800 in the Scottish county of Ayr. Several strains of native cattle are believed to have been crossed with Teeswater stock, which was also used to develop the Shorthorn breed in England. The Scottish breeders developed an animal that was suited to their area's harsh climate. The cows became

Holsteins are the most popular breed because of their ability to produce large quantities of milk. This breed is familiar to many people because of their distinctive black and white markings.

efficient grazers able to survive on the sparse land and less-than-ideal forages. Ayrshires have been valued for their strongly attached, evenly balanced, symmetrically shaped, and quality udders.

Ayrshires are red and white but can vary from white with red spots, red with white spots, all red, or all white bodies. In some cases the red can be very light red, deep cherry red, mahogany, brown, or a combination.

Ayrshires reach a mature size of 1,200 to 1,300 pounds. They are strong, rugged cattle that adapt to all management styles, whether it is group housing or pasture grazing. Ayrshire calves are hardy and easy to raise because of their vitality. They typically do not produce the highest volume of milk, although some herds average 17,000 pounds of milk per cow per year with 3.9 percent butterfat and 3.4 percent protein averages.

BROWN SWISS

The Brown Swiss originated in Switzerland's rough, mountainous regions and is considered the oldest of the purebred dairy breeds. Brown Swiss are known for their stamina and vigor. They are solid brown in color and can vary from fawn to very light brown to dark brown or almost black. Their noses, tongues, and tails are black, as are their hooves. Brown Swiss have large frames and heavy bone structure. This durability suited them well under the rugged conditions of the Swiss Alps region.

Brown Swiss are easily managed, generally quiet cattle that work well in either group housing or on pastures in a grazing program. They can quickly adapt to different environments. They are among the largest dairy breed with cows typically averaging 1,300 to 1,600 pounds at maturity. A Brown Swiss cow produces about 16,500 pounds of milk per year with 4 percent butterfat and 3.6 percent protein averages. Calves grow rapidly and the bull calves grow well as steers.

DUTCH BELTED

One less-prevalent breed that has made a recent resurgence is the Dutch Belted or Lakenvelder. Centuries ago, Dutch Belted cattle were sought by kings and noblemen because of their peculiar and striking markings. Today they still inspire curiosity with their distinctive white belt around the middle of the body with solid-black front and rear ends.

The breed originated in the Tyrol areas of Switzerland and Austria. They were highly prized for their milking and fattening abilities by the Dutch nobility, and the breed began to flourish in Holland. P. T. Barnum imported Dutch Belted cattle to the United States for show purposes. They were later placed on a farm and this appears to be the beginning of the breed in America.

The cows are moderate in size and weigh 900 to 1,500 pounds. Their milk tests range from 3.5 to 5.5 percent butterfat. They are considered an intelligent breed with friendly, quiet dispositions that are valued by families who use rotational grazing or other systems where ease of handling is welcomed.

Dutch Belted cattle have several advantages that may offset the lower production volumes, such as birth weights that average about 70 pounds to ensure calving ease and reduced calf loss. The breed has an excellent grazing ability. Heifers utilize forages efficiently and find favor in grass-based dairy systems.

GUERNSEY

Guernsey cattle originated in France and the Channel Islands and specifically on the Isle of Guernsey. They were introduced into the United States around the year 1840. Guernseys are characterized by their light fawn color with white markings.

For many years this breed vastly outnumbered others in the United States because of its ability to produce high-butterfat milk with a high concentration of beta carotene. Golden Guernsey milk was heavily promoted because of this fact and there still are many people who believe the milk has many beneficial health aspects.

The breed went through a steady decline when fat content became less important and breeds that produced larger volumes of milk became popular. Guernseys are again increasing in number in the United States because of grazing abilities, gentle disposition, calving ease, and efficient milk production.

Guernseys are intermediate-sized cattle with cows averaging about 1,100 pounds at maturity. They produce their milk while consuming 20 to 30 percent less feed per pound of milk than larger breeds and will produce an average of 15,000 pounds of milk per cow per year with 4.5 percent butterfat and 3.6 percent protein.

HOLSTEIN

The Holstein, as it is commonly known today, was originally known as Holstein-Friesian because of its origination in the province of Friesland in the Netherlands and the Holstein region of northern Germany. The cows were bred to make the best use of the landscape in Holland, which was mainly grass. As the Dutch Friesians and Holstein cattle intermingled, their descendants developed into the distinctive black and white colors still prevalent today. Holsteins came to the United States in large importations during the 1860s and 1870s. Some Holsteins possess a recessive gene for red hair color, and red and white animals commonly live alongside their black and white counterparts.

Holsteins are known for their ability to produce large quantities of milk that, in some cases, can approach the

Jerseys have the highest average butterfat and protein content of all dairy breeds, which has led to their popularity. They are excellent grazers and can thrive on many types of pasture conditions.

butterfat and protein content of other breeds. The average Holstein cow produces about 18,000 pounds of milk a year with 3.2 percent protein and 3.6 percent butterfat averages. Holsteins can attain a large mature size with the cows reaching 1,500 to 1,600 pounds. Because of their large frames, they often require larger housing to best suit their size. Calves can be large at birth, which may necessitate more routine assistance. However, heifers grow fast and can usually be bred at 15 months of age when they weigh about 800 pounds.

Holsteins can consume large volumes of forages and can be good grazers because of the amount of grasses they eat. They don't thrive on poor pastures and are not as efficient at converting feed to milk as other breeds.

JERSEY

The Jersey breed originated on the Island of Jersey, a small British island in the English Channel off the coast of France. The breed came to the United States in the 1850s and is one of the oldest dairy breeds with a known ancestry of nearly 600 years.

Jerseys are adaptable to a wide range of climatic and geographical conditions and are found all over the world,

Milking Shorthorns are excellent grazers and have been successfully used in crossbreeding programs, particularly with red-and-white Holsteins. They range from red to white to roan in color markings.

including Denmark, Australia, New Zealand, Canada, South America, South Africa, and Japan. They are more heat-tolerant than larger dairy breeds but smaller in size. The average Jersey cow weighs 800 to 1,200 pounds at maturity. The calves are smaller and birthing difficulties are less prevalent when compared to other breeds. Because of their smaller size, Jerseys can produce more pounds of milk per pound of body weight than any other breed. Some Jerseys can produce 13 times their bodyweight in milk per year.

Jerseys range from light gray to light or dark fawn to black, and from white-spotted to solid in markings. They may have black in the tail switch and the muzzle with a light-colored encircling ring. Jersey milk is about 5.0 percent butterfat and 3.8 percent protein on average. This makes them attractive to farmers in specialized protein marketing programs. The average Jersey cow will produce about 15,000 pounds of milk per year. Jerseys generally have excellent udder shape. They are excellent grazers and have made a tremendous resurgence in the total percentage of all dairy cows in the country. Jerseys can thrive on medium to poor pastures as their maintenance requirements are lower than cows of larger breeds.

Jerseys typically respond to the kind of management they receive. If handled gently, they can become pets and are very docile. Poor management and treatment may cause them to become mean, nervous, or sensitive.

MILKING SHORTHORN

Shorthorn cattle originated in northeastern England in the Valley of the Tees River, in close proximity to Scotland. They were imported to the United States as Milk Breed Shorthorns in 1783 and were sometimes referred to as Durhams. They became favorites for meat, milk, and as a source of power for fieldwork.

Milking Shorthorns are not a separate breed but rather a segment of the Shorthorn breed; their ancestry can be traced back to the same origin. During the 1700s, separate blood lines were developed: one that was leaner with good milking qualities and one that was thicker, blockier, and meatier.

In the past 20 years, the Milking Shorthorn breed has made much progress in transforming a beef/dairy animal into one designed primarily for milk production. Milking Shorthorn cows average about 16,000 pounds of milk per year with 3.7 percent butterfat and 3.4 percent protein averages. The cows can reach 1,100 to 1,300 pounds at maturity.

Milking Shorthorns are red, white, red and white, or roan. They are one of the most versatile breeds and can adapt to a wide range of management systems, climates, and regions of the country. They are excellent grazers that can produce large amounts of milk from home-grown roughages and grass. The calves quickly thrive after they are born. Milking Shorthorn cows are known for regular

A structurally sound cow can last many more productive years than a cow that isn't sound. A thoughtful breeding program that takes a balance of body parts into consideration can produce calves that will grow into sound, useful cows. By knowing the parts of a cow, you will become more familiar with breeding program terminology to make better choices.

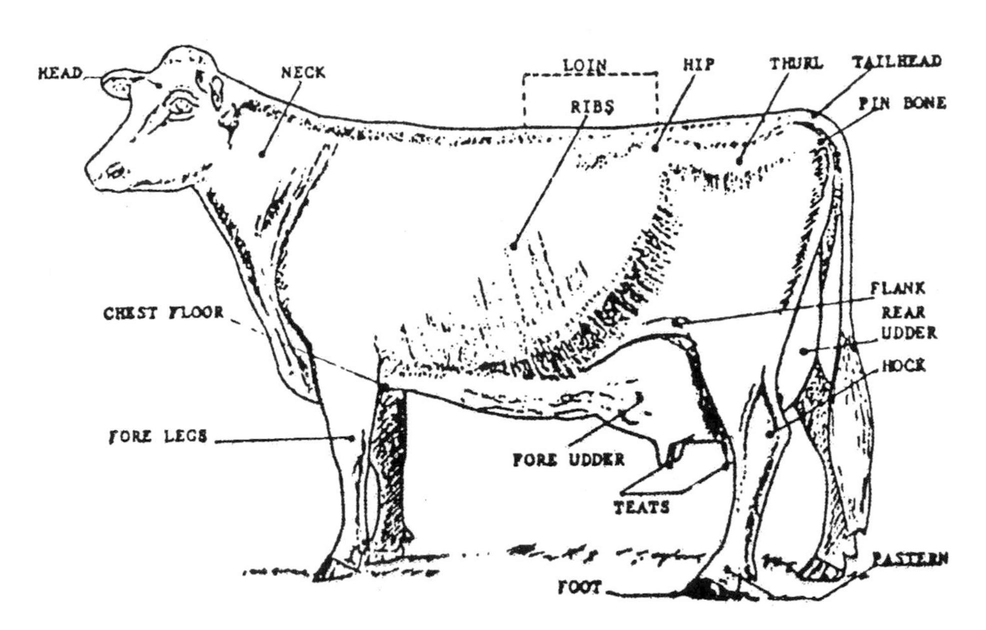

calving intervals, living long productive lives, and having a very desirable quality grading carcass.

CROSSBREDS

Some international dairy breeds are being used in the United States for crossbreeding purposes. Among these are Normande, Norwegian Red, Red Angler, and Aussie Red. It is not possible to import live animals from these breeds but semen from bulls of each of these breeds can be secured for a crossbreeding program. While the performance and adaptability of the dairy cows that you may want for your farm are readily available in breeds already in the United States, some dairy farmers find adding these specialized breeds to their program to be interesting and worthwhile.

STRUCTURALLY BALANCED COWS LEAD LONGER LIVES

Factors such as culling rates, animal health, reproductive efficiency, and calving ease may have as much to do with cattle management as with the breed itself. A structurally sound cow will live longer and produce more milk and calves than one with physical problems such as bad feet and legs, swollen udders, and an inability to stay healthy due to the stress of production. Through careful breeding, cows can be matched with bulls to produce a calf that will grow into a more structurally correct animal.

One of the longest recognized programs available to help dairy farmers develop a successful breeding program is the independent Animal Analysis Associates (aAa). For more than 55 years it has successfully helped breeders produce more physically balanced animals with increased longevity greater milk production, and easier calvings.

Some dairy producers use crossbreeding in their programs by crossing one dairy breed with another. This is an example of a Brown Swiss-Holstein cross. The purpose is similar to beef crossbreeding: You are trying to incorporate the best qualities of both breeds.

CHAPTER 18

COW CARE— MANAGING DAIRY CATTLE

A healthy animal is a productive animal. Providing the best care for your animals will result in better performance. Good care includes proper feeding, care at calving time, daily observation, and good milking techniques. Proper nutrition is the most important element in a profitable dairy operation. Attention to nutrition starts before the calf is born and follows until a heifer becomes a cow. Too often dairy farmers think of young heifers as a cost rather than an investment in their future.

There are costs associated with raising a heifer and developing her into a cow, but the time and feed invested to ensure proper growth to reach target weights is paid back with increased production. Calves neglected or improperly fed exhibit stunted growth and will not achieve good performance levels as cows. It is good stewardship to provide your animals with a sufficient diet so they grow to their fullest potential. Stewardship not only applies to the environment and land but also to animals in your care.

Healthy animals are the most productive animals. By providing proper nutrition, immediately attending to problems, and raising healthy calves you can sustain a herd that produces quality cattle for many years.

UNDERSTANDING CATTLE BEHAVIOR

A dairy cow's productivity is directly related to the care she is given. Cows respond positively or negatively to comfort or discomfort and good treatment or maltreatment.

The best and most alert handlers of dairy cows have the perception and ability to read body language in animals. By watching the movements of their cows, these handlers can instinctively tell whether the cow is comfortable. If the cow isn't comfortable, the handler will take steps to improve the situation. Sometimes the way animals behave is the only clue that stress is present.

Bellowing, kicking, and general agitation are signs that a situation needs your immediate attention. A cow will exhibit signs of illness by her head or ears drooping, her tail being listless and uninterested in swatting flies, and by the fact that she is wanting to lie down constantly. Cold ears, when touched, should alert you that she may be getting sick.

The normal physical processes of the dairy cow require a body temperature that is maintained within very narrow limits, between 100 to 103 degrees Fahrenheit. The normal core body temperature of a healthy, resting cow is about 101.5 degrees. Although every cow's metabolism is different, this normal temperature will not vary much. Anything outside of this range should have your attention.

The cow's environment has an effect on her body temperature, as well as the time of day and how active she is. Her temperature will be slightly lower in the morning, due to the rest she received, and higher in the evening after a day of activity. A cow's body temperature will also be affected by the climate and weather conditions where she lives. Humidity, wind, and the sun interact with the air temperature and affect the cow's ability to maintain a

Calves can be raised inside buildings or outside in hutches made of wood or plastic. The calves are fed individually until they are old enough to be put together for group housing in larger pens or corrals. These calves are your future herd so you need to provide them with adequate facilities and proper nutrition.

steady temperature. As the air temperature rises, her stress increases as her body works to expel the excess heat. Providing some form of shade or shelter will help so she can stay out of the direct rays of the sun.

Humidity is probably a greater detriment to a dairy cow than hot sunlight. Shade can counteract the sun's rays, but when humidity levels are high there are few ways to keep her cool. Some dairies install overhead water sprinkling systems that spray a fine mist on the cows as they stand in the shade to help cool body temperatures. Lack of appetite is one result of high body temperatures, and this results in lower milk production.

In extreme cases the cow's core body temperature can rise to dangerous levels. When this happens, immediate action must to be taken to cool her down or else heat stroke can occur. In times of extreme heat, cows need full access to water. Having an alternative water source will allow many cows to drink at once if the dominant cows control access to the water.

Many cows have calm, docile personalities that make them enjoyable to work with every day. Their personalities make them as unique and individual as any human.

THE DAIRY BULL

Because of artificial insemination, it is no longer necessary for farmers to keep bulls on their property. However, some farmers will use bulls to breed cows that have difficulty becoming pregnant through AI. Some dairy producers use gomar bulls that have been castrated to help them identify which cows are in heat and need to be bred. These cows are then sorted out for artificial insemination. Having a dairy bull of any size on your farm is a risky proposition and one that you should consider very carefully, particularly in relation to your family if you have young children.

Because of their size and disposition, bulls must be considered one of the most dangerous domestic animals. There are many instances where an owner or family member thought a bull was safe, friendly, and quiet only to have him turn on the individual in an unexpected moment and cause him or her serious injury or death.

If you choose to keep a bull on your farm, whether in a pen or in the pasture with the cows, be aware of some of the body postures he may exhibit so that you will safely know how to react. By staying ahead of the bull mentally and physically, you will have a margin of safety.

The cow herd is a bull's domain and he perceives anyone trespassing on his territory as a threat and will display postures of aggression such as putting his head down, displaying a broadside body view, bellowing, pawing the ground with his feet, rubbing the ground with his head, shaking his head from side to side, or advancing toward the object of his discontent. These challenges or displays of territorial domination need to be taken seriously. If you and the bull are in a corral, your chances of escape are greater than if this happens in an open pasture because he is fast enough to catch you.

When these threat displays are in progress, the best thing to do is quickly move backward and get to a safe spot, whether it is behind a gate, fence, or nearby machinery. Look for anything that will provide a barrier between you and the bull.

Always knowing your position relative to the bull will help you avoid serious consequences. One rule to follow in all cases it is to never turn your back on a bull. Always keep him in front of you in full vision. It is wise to be wary of

Cows need shade during high summer temperatures to stay cool, especially if they spend many hours on pasture. If there is no shade available, they will bunch together and increase the problem. Maintaining milk production in hot weather can be a challenge, but with adequate shelter or shade the effects from heat can be reduced.

any dairy bull and understand that they are not to be trusted under any circumstances.

One word of caution. If the bull on your farm has attacked you or anyone else, it should be his last attack and he should be immediately sent to slaughter. Once a bull has crossed that line, he knows he is in command and the situation will never be the same between a bull and his owner.

Some recently fresh cows can also display a threatening posture, but their actions tend to be more subtle. Although they are not as great a problem as bulls, some cows may display aggressive maternal protection for their calves and can become a serious threat to you.

Use caution when approaching a cow with a newborn calf. The cow will tell you if she wishes to share it with you or not. Gentle mooing will tell you that she is comfortable with the situation. Approaching her where she has you in full view will keep the cow from being surprised. Talking in a low, gentle voice will help her relax and may relieve her apprehensions since she is already familiar with you.

If she starts to shake her head or tries to lunge at you, it is best to back away and become less threatening to her and the calf. She is only trying to protect what is hers and she is not going to share the calf with you.

Studying cattle and their behavior will help raise your knowledge of each animal and will improve your husbandry skills. By understanding patterns of normal and abnormal behavior in animals, you can improve the care and handling of your cattle. This will help you achieve better animal comfort, welfare, and safety.

COMMON DAIRY PRODUCTION DISORDERS

There are a number of common diseases and disorders that affect dairy cattle at various stages of life. Calves are affected mostly by diarrhea and pneumonia, which are the leading causes of death in calves. Older heifers can be afflicted with pneumonia, injury, bloat, pinkeye, and other minor conditions. Cows can have a number of disorders that result in lost milk production. These can be categorized into metabolic, environmental, structural, or reproductive disorders.

METABOLIC DISORDERS

Metabolic disorders include milk fever, ketosis, fatty liver syndrome, and grass tetany.

Milk fever (hypocalcemia) is not a fever related to temperature, although the body temperature of a cow is generally depressed if she has this condition. This condition occurs at calving time and is caused by a sudden shortage of blood calcium. The signs are staggering, difficulty in standing, and not being able to get up. Immediate intravenous calcium treatment is required or the cow may die. A proper dry cow diet can prevent most cases from occurring.

Ketosis (acetonemia) is most frequently observed in well-conditioned cows from two to six weeks after calving. It occurs when a cow's liver has been depleted of stored glycogen. Energy from fat is mobilized to the liver and produces ketone bodies that are burned throughout her body. A rapid utilization of body reserves and insufficient or impaired carbohydrate usage will cause cows to drop in milk production in only a few days. Administering propylene glycol or sodium propionate as an oral drench should quickly relieve the conditions. Test kits for milk and urine are available to detect this condition early. Limiting grain amounts after calving will help avoid this condition.

Ketosis and milk fever can be successfully treated if caught in time. Propylene glycol or sodium propionate is used as an oral drench for treating ketosis. Calcium is the best treatment for milk fever and is given intravenously. These products are available from farm supply stores or your local veterinarian.

Fatty liver syndrome is the accumulation of fat in the liver. Dairy cows do not normally store fat in their livers. This condition occurs when a cow breaks down more fat than her liver can properly process. Signs of fatty liver syndrome include a reduced appetite and milk yield. When the case is severe, it can cause death. With the liver not working to its normal level, other body functions are affected. Secondary conditions may occur, including ketosis and a displaced abomasum. Cows that are fat at calving are more susceptible to a fatty liver condition. Treatment includes glucose, propylene glycol, and corticosteroids. Prevention is more important because the treatment process can be long and is often ineffective. Feeding a balanced dry cow ration and preventing excessive weight gain during the dry period will help.

Grass tetany (hypomagnesium) is usually observed in cows that graze on lush grass pastures high in nitrogen, which results in low absorption of magnesium. The signs include walking with a stiff gait, falling, going into convulsions, and then death. This condition most generally occurs when cows are grazing very young grass pastures since they have a lower magnesium level than older grasses. Heavy nitrogen use on young pastures reduces the levels and amount of magnesium available to the plant from the soil. Prevention is a better option than providing treatment, which is usually done with an injection of calcium and magnesium intravenously followed by an injection of magnesium under the skin. The best prevention is to watch your cattle when pasturing them on grass fields that have been heavily fertilized with nitrogen. Using a supplement of two ounces of magnesium oxide daily during the danger period should provide enough help to avoid this condition.

ENVIRONMENTAL DISORDERS

Your cow's environment can cause several different disorders, diseases, and poisonings including chemical poisoning, plant poisoning, toxicities caused by moldy feed or urea, acidosis, hardware ingestion, displaced abomasums, and mastitis.

Chemical poisoning is not a problem if you farm organically because you won't have stored crop chemicals. However, chemical poisoning can occur on farms subject to drift from aerial spraying or direct spraying on adjoining farms. Your farm may have a stream or other types of waterway running through it where the water originates on a farm where chemicals are used. The chemicals may travel with the water in times of heavy rainfall or snowmelt and your cattle can be exposed. Poisoning may result from exposure to insecticides, pesticides, herbicides, or parasiticides. The best prevention is to be aware of these conditions near your farm. Being cautious about crops that are harvested after chemical application, reading and understanding labels, and following directions are the best preventative measures. Call the vet right away if you suspect your cow has been poisoned.

Plant poisonings generally occur when toxicity is built up due to abnormal harvesting conditions. Nitrate poisoning occurs when blood hemoglobin cannot carry oxygen to the body cells due to an excessive intake of nitrates or nitrites. Weeds and plants stressed from frost or drought are the most dangerous. Common poisonous plants include bracken fern and nightshade. When feed is scarce, cattle will eat whatever is available and may consume enough plants to produce a toxic effect. Prussic acid poisoning occurs when drought- or frost-stressed sorghums or sudan grass produce toxic hydrocyanic acid. Avoid feeding or letting cattle graze on young plants less than 24 inches tall. Feed dry hay to reduce the risk.

Moldy feed toxicity occurs when a fungus or other mold grows in feed grains that are stored in moist conditions with poor ventilation. The easiest prevention is not to feed any grains, hay, or silage that has mold in it. Urea poisoning occurs when insufficient carbohydrate intake results in excessive ammonia in the rumen or too much urea is fed at one time. Keeping the amount of urea or ammonia supplement to under a quarter pound per day per cow will prevent problems.

Acidosis occurs when the rumen develops an acidic condition with a pH of 4.0 to 4.5. This impairs rumen function and digestion, resulting in a loss of appetite and dull appearance. Avoid having accidental access to grain,

If using chemicals on your farm, be sure to take precautions that they are used and stored properly so that cattle and children cannot gain access to them. In undiluted states, chemicals are very potent and can have an immediate harmful effect if swallowed.

grain carts, feed rooms, and feed bags where cattle can rapidly ingest large amounts of grain. Also, avoid rapid changes in a feeding program to any high-energy diet, such as grain.

Hardware ingestion is not so much a disease or disorder as it is a condition from something the cow swallowed, usually involving metal. Hardware ingestion results when a sharp object punctures the reticulum and the animal suddenly loses its appetite, has a reluctance to move, or moves gingerly. Avoid pasturing and making hay or silage from field areas that may have contained old fences where wire may be present. You can avoid many hardware problems by orally giving bolus-sized magnets to your cows. These magnets lie in the reticulum and move around in the churning material and gathers bits of metal and wire that the cow may have swallowed and keeps it from puncturing the stomach wall or heart.

Displaced abomasum is when the animal's fourth stomach moves in the body cavity (generally after calving), twists, and prevents feedstuffs to pass to the rest of the digestive system. You or a veterinarian will most likely diagnose it by listening to a pinging sound when using a stethoscope. Surgery is required to relieve this problem. The operation involves suturing part of the stomach wall to the inside of the cow's body cavity to prevent a displacement from occurring again. Feeding bulky dry hay prior to and at calving time will help reduce the incidence of a displacement.

Mastitis is an inflammation of the udder tissue caused by one of several bacterial agents, most commonly staphylococcus or streptococcus organisms. The environment has the largest influence on mastitis infections due to improper milking procedures, unclean bedding, stress from heavy production, weather conditions affecting barn lots, and a host of other reasons. Inflammations can be minor or so severe that gangrene sets in. Good hygiene, proper milking techniques, and clean barns are the most important measures you can take to reduce the incidence and severity of mastitis cases. If a cow develops mastitis, call the veterinarian to discuss treatment options.

STRUCTURAL DISORDERS

Some physical problems originate because of the structural makeup of the cow. These include foot rot, hairy warts, and udder edema.

Foot rot is a condition due partly to physical structure and partly to environmental conditions. It is a break in the

Cattle will seek shade during hot weather. Providing access to a pasture with shade is important in keeping body temperatures within a normal range and reducing heat-related problems.

skin or hoof either between the toes or on the heel of the foot. When a break occurs, bacteria in dirt, manure, or the farmyard can enter these cracks and cause swellings. Progressive lameness is the most obvious symptom, and in extreme cases, the cow will have a swollen foot. Infections can progress into the joints and spread into the leg or bloodstream. Having clean yards and facilities free of sharp materials that can break the skin or hoof will help lessen the chance for infection. The cow's hoof will need to be cleaned, disinfected, and wrapped to treat foot rot.

A secondary condition affecting cow feet is hairy warts, which are lesions found on the heels of the feet. In later stages there is a hairy-like protrusion. Two or more microorganisms may work together to cause this condition. For the infection to start the feet must be subjected to prolonged exposure to moisture and exclusion of air. Foot baths using copper sulfate have been found to be very beneficial in treating this condition.

Udder edema is an excessive accumulation of fluid in the udder and usually occurs at calving time and shortly after. It is usually more severe in high-producing or first-lactation cows. Some breeders believe it is more related to the physical structure of the cow than any metabolic or environmental factor. Because it takes 400 pounds of blood pumped through the cow's udder to produce a pound of milk, the heart works extremely hard pumping blood and maintaining the health of the cow. Animals without the heart or lung capacity to handle the stress of the high production will have a hard time moving that quantity of blood efficiently through her udder to flush out the fluids and swelling that accompanies calving. The constitutional strength of the cow will determine how well she moves these fluids out of her udder, while secreting milk at the same time. Stimulating circulation by massaging the udder and using diuretics may help relieve some of the symptoms but generally these are only temporary effects. Time is the best healer for udder edema.

REPRODUCTIVE DISORDERS

There are several conditions relating to reproduction that can affect the health of your cows. Metritis and endometritis are inflammations of the uterus caused by

Bulls in a pasture with cows and calves may make a nice picture but there is a sense of danger involved. Bulls threatened by outsiders can quickly transform from docile to aggressive, and you need to take safety precautions any time you are near them.

The beef bull should be handled carefully, and attention always needs to be paid to his location if you have to enter the corral or pasture where he is. Anticipating possible confrontations will make you aware of his presence and provide a margin of safety.

bacteria, protozoa, fungi, or viruses that generally enter the cow's reproductive system at calving time. After calving, the uterus closes itself around the sphincter muscle. When closed, it retains any bacteria that may have entered at calving. This closed environment is an incubator for bacteria, which usually cause infections to the reproductive system.

These are not the only reproductive disorders or diseases that can occur. A clean calving pen or corral is the best preventative measure you can take. Using a routine reproductive examination program with your local veterinarian will help diminish many problems.

JOHNNE'S DISEASE

Johne's disease is a contagious, chronic, and usually fatal infection that primarily affects the small intestine of ruminants. It is caused by Mycobacterium paratuberculosis, a hardy bacteria related to the agents of leprosy and tuberculosis. It is characterized by persistent diarrhea, gradual weight loss, and general weakness, although the cow may appear to have a normal appetite.

Johne's is an insidious disease because the signs of infection are rarely evident until the animal is two or more years old. The infection usually occurs shortly after birth and calves are most susceptible to the infections during their first year of life. Newborns most often become infected by swallowing small amounts of infected manure from the area where it is born or from nursing on an infected mother.

Some vaccines show promise but once the animal is infected, it is difficult to treat. Providing a clean calving environment and getting rid of infected animals appear to be the best ways of controlling the incidence of new infections.

The dominant cattle of a herd will be first for many things, including the first to access water and feed. Providing several places where cows can get water or feed will allow them all to have sufficient access.

Tests are available that can identify animals harboring an infection of Johne's. If purchasing animals for your herd, you should insist on a negative test for Johne's on each animal before they enter your herd or buy from a herd identified as low risk for Johne's. More information on diagnosing, preventing, and controlling Johne's disease is available from your herd veterinarian.

EUTHANASIA AND THE DEAD BODY

There comes a time on every dairy farm when an animal suffers an injury, illness, or some other debilitating condition that requires euthanasia to provide a swift and humane death. This should be done in a manner that will minimize any stress and anxiety experienced by the animal prior to unconsciousness.

Correct euthanasia procedures will produce rapid unconsciousness, followed by cardiac arrest and total loss of brain function. There are several methods to accomplish this, including using chemicals, which is usually accomplished by a licensed veterinarian who uses a barbiturate product. Animals euthanized in this manner should not be used for human consumption or fed to other animals.

A physical method that does not require human contact with the animal is shooting it with a gun. Strict firearm safety must be observed, as well as local laws and ordinances relating to the discharge of a firearm.

After all vital signs of an euthanized animal disappear, the body must be removed. Most states have regulations relating to the disposal of dead animals. Generally speaking, burials are allowed in most states but they must conform to certain time limits, distances from wells, adjoining properties, waterways or streams and lakes, and residences to name a few. Information on ordinances in your area or state can be obtained from your state department of natural resources or your county agricultural extension office.

Private companies that specialize in dead animal removal is another option. This service may or may not be fee-based, but they are capable of handling large carcasses.

CHAPTER 19

EXIT STRATEGIES— HOW TO END A BEEF OR DAIRY ENTERPRISE

Change happens in life and on farms. For anyone living on a beef or dairy farm there comes a time when a sale occurs, whether it is a complete sale precipitating an exit from farming or a partial sale of the assets (such as the cattle or machinery) and retention of the land. It may be a sale to another member of the family, or be a combination of these instances. As with any other aspect of your business, planning ahead for a sale will generally pay greater dividends than if you let events dictate your course of action.

The reality is that a sale is likely to occur in your lifetime. There is no disgrace in this situation. Farms have been bought and sold for generations and many people have entered and exited farming. That may be the cold reality but there is little doubt that leaving a farm can be an emotional time for you and your family. The reasons for

At some point your beef or dairy program will come to an end. Acknowledging this at the beginning of your pursuit of a cattle program will make this decision easier. Selling your animals does not necessarily mean leaving the farm. Other options may be available at that point that allow you to continue your dream of country living.

leaving may include health issues, finances, changes in a family situation, or a desire to pursue other interests.

In many cases it may seem easier to get into a business than to get out. If you plan to exit from your beef or dairy business, there are basic similarities and differences to any other business that has an exit strategy. A discussion with your financial advisor may provide answers to help you successfully exit.

DON'T FEEL GUILTY

You do not have to feel guilty that you are leaving the farm or having a sale. Too often families who leave their farms develop a sense of guilt because they consider it a failure on their part. Perhaps financial problems contributed to the sale but the fact that they spent time and effort trying to make their business succeed should not be viewed as a failure. Farming can be a challenge even in the best of times. Market forces have a way of blowing down the best-built house, and those who can withstand such events are sometimes more lucky than good managers.

SALE OPTIONS

Farms are different from many other businesses in that they have living assets—animals—that must be sold. They are similar to other businesses because they also have non-living assets to be sold and tax considerations after a sale. There are several ways to handle the sale of your cattle or machinery, and each has advantages and disadvantages. A thorough understanding of each option can save you money and minimize surprises.

PUBLIC AUCTIONS

Public auctions have been a traditional venue for dispersing the assets of a farm. This type of auction allows anyone to bid on the items being sold. In many ways your exit may resemble the same route you used to enter.

One venue where you can sell animals is through a traditional market, such as an auction yard where buyers gather to view large numbers of animals at one time and buy their choice. This may not be to your advantage if few buyers are in attendance to compete for your animals on the day they are sold. There are also costs involved with trucking the animals to the auction site if you do not do the hauling yourself. Other factors that may be involved include the shrink of the animals that occurs during hauling; the risk of injury to any of the animals, which may lower their value at auction time; and the buying atmosphere during the auction, which may have a positive impact if the selling atmosphere is high or a negative impact if the buyer attitudes that day are low and the commissions you will have to pay for this service are high.

You can manage your own public sale but there are several reasons that it is generally better to hire an auction service than to handle your own sale. Auction companies have the experience to provide a quality service. They are bonded, which means they can handle the financial transactions that occur at a sale with a guarantee for assuring that you will receive the money paid for your assets. They have experience to know the audience that will be most interested in your assets and they will advertise in publications to target that audience.

Most auction services have personnel who can help with set up and take down of the equipment needed on sale day. They can advise you about the time of year that would be best to sell your animals and equipment and how to structure your sale. There are expenses with hiring an auction service but many owners feel that the percentage of the total dollar volume of their sale is well worth the price.

If you contact an auction service, be sure to read the contract thoroughly and understand all of the clauses involved in your sale. If you are unsure about anything, ask or have your lawyer or financial advisor read the contract before you sign. While most auction services are reputable businesses, it is simply good business to be certain about the conditions of your sale and the costs.

An advantage of having a public auction is that it creates competition among buyers and the final price may exceed that of a negotiated sale. It takes the owner out of the negotiation process and you know exactly when everything will sell. One disadvantage of a public auction is that the animals or equipment sell to the highest bidder regardless if you think it is a fair price or not. Unless it is a reserve auction where you can put a minimum price on each item, you are obligated to sell at that highest bid.

PRIVATE SALES

Private sales are when you sell directly to another party. This is typically a situation where you advertise your animals or equipment for sale in a trade or farm publication and

Having a public auction is one way to sell your cattle and may be the venue you used to enter cattle-raising yourself. One key to a successful public auction is to plan far enough ahead so that all details are handled at a time when there is little pressure on you. This may include pregnancy checks, vaccination scheduling, advertising, and promotion.

wait until someone contacts you. Word-of-mouth is another way to announce the sale of your assets. With a private sale you set the price and negotiate with those who wish to purchase your animals or machinery. This provides an opportunity for potential buyers to come to your farm to view your assets.

One of the disadvantages of this type of sale is that you will be responsible for making certain that the payment is received and substantiated prior to any animals or equipment leaving your farm. Also, there may be a time lapse between the time you announce you want to sell your assets and the final sale. This may be dependent upon the market conditions at the time, the interest level of the public, or how many potential buyers see the ad.

The owner is responsible for all negotiations on the price for private sales and this may be a disadvantage if you do not feel comfortable with negotiations or it may be to your advantage if you have a talent for bargaining.

DEALER SALES

Dealer sales are another option in selling your animals. A dealer is someone who buys animals for himself or for other people. In some cases the dealer will handle the transaction between private parties for a commission fee based on a per head or total volume sale. You will be best served by working with someone whose good reputation is recognized by your neighbors and friends. Ask them for recommendations of cattle dealers whom they would trust to handle a transaction.

There are still many transactions that take place on farms with a simple handshake. If you have known certain auctioneers or dealers for a long time, this may be a situation in which you feel comfortable. However, to protect yourself, get a written contract so all expectations and requirements are understood and will eliminate any misunderstanding should something go wrong.

OTHER OPTIONS

Video auctions and Internet auctions are additional options for selling animals or other assets. These are typically more useful when selling a large number of animals and may not fit your situation but, as with other new forms of marketing, they may hold promise for the future and may be a consideration for you in future years.

Selling your animals to another party through a private transaction is another way to sell your cattle. It involves less organization but will require more print advertising or word-of-mouth exposure to ensure that potential buyers know your animals are available. Some salesmanship on your part may be required, but involving a reputable cattle dealer in the negotiations may be helpful to ensure a successful sale.

POST SALE

The conditions for having your sale may determine what happens next on your farm. If you sold all the animals from your farm, then the land will still be part of your assets. If you sold a dairy herd, the options for your farm could include raising beef cattle. With the facilities available, the transition from a dairy farm to a beef program is relatively easy. If you have been dairying, chances are there is good fencing available to handle beef cattle.

Your facilities can stay empty if your animals are gone but this will mean paying taxes on unused buildings. Your buildings can be rented to another party to provide income. A written contract for renting to another party is good business and will keep any potential problems to a minimum. Additionally you can rent out your land while still retaining ownership and living on the farm. Another option for your farm if you discontinue raising cattle is to convert your land into a vegetable produce farm or other cash crops.

POSTSCRIPT

From these examples it should be obvious that the sale of your cattle does not need to be the end of your farming life unless you choose it to be. There have been many farmers who enter a cattle-raising business, leave, come back in, and then leave again. Some try to avoid the peaks and valleys of the cattle marketing cycle and use an entry and exit policy.

Another reason for selling and going back into raising cattle may have to do with wanting time away from the farm for health and personal issues. Selling your cattle does not have to be a traumatic experience if you accept the idea that a sale will happen eventually either with or without you in attendance. Planning ahead for a sale can take a lot of the emotional stress out of the decision. It can leave you with a healthier state of mind and a satisfaction that you have accomplished the goals you set at the time you entered the beef or dairy business.

CHAPTER 20

COUNTRY LIVING—REALITIES OF RURAL RESIDENCY

Requesting visitors or others entering your farm to use disposable plastic slip-on boots is an easy way to ensure that no foreign matter or manure from another source is brought to your herd. These precautions may seem excessive but can help maintain a healthy herd.

People who own a farm in the country and people who own a house in the country have two different perceptions of the same area with a wide gulf in between. The purpose of your farm is to produce agriculture and the purpose of a rural home is to enjoy all the benefits of living in the wide open spaces without having to contend with many urban issues including traffic, close neighbors, and noise at night.

With two different agendas at work in the same area it is understandable that conflicts and disagreements can arise as to which agenda takes precedence. If conflicts arise between farmers and new rural neighbors it is usually over such things as smell from hauling manure, dust while working fields in the spring, or noise if harvesting is done at night. There are many other things that can cause grievances between these two groups of people but these are the issues that seem to provide the most misunderstanding.

Many states now give farmers a basic right to farm without the fear of lawsuits brought by offended neighbors. In these cases an agricultural operation is presumed not to be a nuisance to the neighbors even when new neighbors move in. If the farm operations are conducted in a reasonable manner, the new neighbors can't legally complain. Landowners, residents, and visitors must be prepared to accept the effects of agriculture and rural living as normal and understand that they are likely to encounter a number of practices that area farmers have been and likely will continue doing in their normal farming practices.

However, there are different conditions attached to these ideas in some states because the right-to-farm laws do not give farmers complete freedom to do as they please. In these instances, farmers must operate in a legal and reasonable manner to be eligible for the law's protection. Some states have developed a list of specific annoyances that are not considered a legal nuisance to neighbors including odor, noise, dust, and the use of pesticides or other chemicals, the very conditions which, without the laws, could lead to a lawsuit by a neighbor.

As more urbanites seek residence in the openness of the countryside, the potential for problems and misunderstandings increases. Opening a discussion with new neighbors can defuse situations when they do not have an understanding of typical farming practices. A proactive approach of reaching out to them instead of resenting their presence or ignoring them may be in your long-term best interest.

IN YOUR BACKYARD

It is often impossible to stop people from purchasing land near or adjacent to your farm. You have little control over what their expectations are for moving into the neighborhood. You may feel they should be grateful for having the chance to live where you do, just as you have decided to venture into farming for possibly similar reasons. But that may not be the position they take and they may feel it is their right to be there without the inconveniences agricultural production imposes upon them. You may find yourself confronted with disenchanted neighbors over something minor, perceived or otherwise, but there are ways of disarming any confrontational situation before it gets out of hand. Taking a proactive approach before it reaches this stage may be your best defense.

BUILDING BRIDGES

Many conflicts can be solved and some can be avoided altogether by using strategies to head off potential conflict and build stronger ties with your nonfarm neighbors and with your local community. Perhaps the best way to avoid conflict is to use responsible farm management practices that contribute to the best environment for everyone. Protection of groundwater supplies, controlling odors when possible, and keeping weeds from becoming a problem are some practices that can highlight your commitment to a good relationship. Besides being good stewardship practices, these will be your allies in the case where you have a neighbor disagreeable to anything you do and who prefers a legal confrontation.

Unexpected new neighbors are an important reason for having good fences around the perimeter of your farm. It is not acceptable to let your animals wander and roam onto another's property by assuming that since you were there first you do not need to keep good fences. Similarly, the fences are a visual boundary for your neighbor and a reminder they must also respect your property and not roam on your land without permission.

Farmers' concerns about keeping infections and viruses from entering their farms and herds have increased in recent years. This includes placing newly purchased animals in quarantine to ensure they are not carrying in any diseases. Blood tests, physical examinations, and visual inspections should all be performed prior to mixing the new animals in with your herd.

Talking with your new neighbors is another way of diminishing potential problems. Simple gestures like giving them a farm tour, hosting picnics, giving away free manure, and perhaps providing garden space in a corner of a field that is not farmed could be a way of reaching across the divide.

The purpose is to create a stronger sense of community, which may be one of the reasons that brought you to your farm in the first place. Don't lose sight of that attraction because it can have a similar hold on other people.

FARM VISITS

One step you can initiate is to invite your new neighbors to visit your farm and see it up close. A friendly atmosphere where you can discuss your farming practices, your procedures, the goals of your business, the values you hold for the land, and the reasons for involving your family in the farm may help offset their concerns as they witness your family's total involvement.

Familiarizing them with the reasons you chose to live in the country and work with animals may give them a better sense that sometimes allowances will need to be made on their part so that your farming practices can continue unimpeded. Explaining how the work on your farm proceeds may be helpful if they have little or no knowledge about the seasonal requirements of farming. There may be times when field work is done at odd hours of the night or day. Helping them understand that these hours occur only sporadically and usually during planting and harvesting may ease their concerns.

You may find that one reason your neighbors moved to a rural area is for the health of a child, especially one that may have respiratory problems. Developing a sense of being neighbors and knowing of conditions where you may be able to prevent distress within their family is a good way of avoiding conflict. By knowing these conditions it will be easier for you to inform them of your intentions for field work or harvesting near their home several days ahead of time so that they can make adjustments to their schedules and routines and can remove their children from the area where you will be working.

Although you may not feel that it is your position to make the first contact, etiquette and politeness are always in fashion, especially toward your neighbors. Working through conflict involves finding common ground and shared interests. In this case, the shared interest is the desire of farmers and non-farmers, whether new or long-time residents, to enjoy the kind of life that the countryside has to offer. As farm families continue to have more new neighbors, building bridges will be vital to working through these challenges. Never underestimate the power of communication to avoid or solve problems before they arrive. When a problem or conflict does present itself, you can feel confident that a workable solution can be made.

BIOSECURITY CONCERNS

The communication and personal contact with your neighbors must be balanced with the need for biosecurity on your farm regardless of size. In the context of livestock production, biosecurity refers to those measures taken to keep disease agents out of your herd where they do not already exist. These measures exist on three levels: national, state, and individual herds.

Biosecurity should not be confused with bioterrorism or agro-terrorism because they are not the same thing, although biosecurity is a part of each concern. As a farm owner, the responsibility for the biosecurity on your farm rests with you, as well as the plans and steps taken to prevent the introduction of any infectious disease onto your farm and to limit the spread of any disease already present within your herd. To be successful, your plans must address how infected animals will be isolated away from other animals in the herd and how cleaning and disinfecting procedures will be used.

The greatest risk of introducing an infectious disease onto your farm is by bringing in new animals that have been exposed to a disease or have it themselves. This is one reason that visual observation of any new animals is important as the first line of defense being the initial screening of the animal.

There are tests available to determine if dairy cows have streptococcus agalactiae mastitis in their udder, whether beef and dairy cows test positive for Johnne's disease, or if beef cows test positive for bovine viral diarrhea (BVD) and other diseases. Insisting upon and making use of these tests can prevent problems from occurring where none exist. A discussion with your local veterinarian about isolation procedures for newly purchased animals or animals already infected in your herd will give you a plan to implement before it is needed. Your veterinarian can also help develop a vaccination program and disinfection plan that will raise the level of resistance to infectious diseases in your herd.

Your concern over biosecurity does not need to impair your relationship with your neighbors. Most established farmers are already aware of these concerns and some may have procedures already in place. Common sense should help dictate your approach to biosecurity. Having visitors is a good way to expose them to your farm, but asking them to wear protective boot or shoe coverage will subtly show that you take the health of your animals seriously. It is rare to have a visitor who refuses your request to put on a pair of disposable plastic slipover shoe covers before entering your barn or farmyard.

DEFENDING FARM PRACTICES

Animal rights campaigns occasionally reach front page news and are a subject with which every farmer must take time to familiarize himself or herself. By giving humane care to their animals, farmers for generations have known that it is in their best interests and the comfort and thriving nature of their animals to provide them with certain unspoken rights such as food, shelter, light, and clean barns.

This stewardship has been appropriated in the last few decades by groups whose agenda has included the elimination of animal agriculture. Considering that only 1.6 percent of the United States population live on farms today, there is no question that agriculture is getting smaller and urbanization is growing.

Animal rights have become a social issue rather than a scientific one. As these views spread and gain traction, it is up to livestock producers to address these perceptions and defend their practices not just scientifically but morally as well. It is about what the animals represent and the values associated with raising them.

Training kits are available to help you and your family understand the issues and how to respond to those who do not from various farm organizations and livestock groups. Raising animals that provide food for other people is noble work and one that has given great satisfaction to many farm families.

Disposable boots can be deposited in marked waste containers to eliminate the possibility of contamination on your farm or visitors taking anything from your farm to their homes.

CHAPTER 21

HELP AND ADVICE— RESOURCES AVAILABLE TO YOU

You will have questions from time to time about how to handle certain problems that arise on your farm. Help is available to assist you in getting answers to your questions, solutions to your problems, and new perspectives to consider.

COUNTY AGRICULTURAL EXTENSION SERVICES

The agricultural extension service, usually located at your county seat, can provide help with answers and solutions. Extension educators are specially trained and have access to their state's university systems and research departments. They are able to glean a large amount of information from these sources and pass it along to you.

Because of their extensive contacts across the country and, increasingly, around the world, extension educators receive information on traditional practices as well as the latest innovations to arrive on the scene as new approaches make front page news in farm publications. Although you may have the same access to electronic data, their background and experience can help narrow your focus, and their services are free.

TECHNICAL SCHOOLS

Technical schools provide information, training, and assistance by offering a hands-on approach. Classes offered by technical schools can supplement and provide training that may not be available through county agriculture extension offices. Their classes may help you prepare other areas of your farming business not directly related to cattle, such as accounting or financial assessments.

GRAZING NETWORKS

Grazing networks are an excellent way to get to know other farmers in your area or region who are developing grazing programs. These are informal groups with the purpose of helping beginning grazers solve problems, explore alternative methods, gain social contacts, and generally have fun while learning about the ways other grazers handle their pastures, problems, and successes. Because these networks are made up of like-minded people, ranging from experienced to novice, their attitudes reflect helping one another rather than forcing views upon new members.

Grazing networks typically have a membership mix of livestock farmers with dairy, cow-calf, beef, and sheep owners among the most prevalent. The main focus of the grazing network is the pasture walk, which typically takes place during the main grazing season from April to October.

Typically pasture walks are held on a different farm each month and begin with a brief introduction by the host farmer or extension educator with a short discussion prior to going into a pasture to walk and observe the management scheme that the host farm is using. Questions are encouraged. Topics covered can include pasture composition or what grasses are planted and being used for forage, fencing layout, how that particular system works for the owner, watering systems, lanes, and the overall grazing management being used. Some networks use a series of walks on the same farm during the year to follow the farm's progress and learn how the host farmer handles the changes over the grazing season.

UNIVERSITY SYSTEMS

All land-grant colleges in the United States provide agricultural classes that are available to the public. The

university system can provide research data on a variety of subjects. These reports may help in making decisions related to your farming practices.

At one time the independent status of university research was unquestioned. However, the increasing budget constraints of universities have caused them to seek funding in some cases through private companies and foundations.

The private nature of these companies and foundations has caused some concern about the ability of the universities to remain unbiased when the finances are provided by companies with a profit motive. The research may not be tainted but whether the wrong questions are being asked or the right questions are not being asked may be a result of losing financial support and should make you closely examine new studies being promoted.

STATE DEPARTMENTS OF AGRICULTURE

The United States Department of Agriculture (USDA) is the cabinet level agency that oversees the vast national agricultural sector. Its duties range from research to food safety to land stewardship. Every state in the country has a department of agriculture that administers the programs of that particular state and operates under its statutes. These agencies have a wide range of booklets, pamphlets, and other publications available to help you understand the rules and laws, obtain licenses, and comply with regulations pertaining to your farming business.

Grazing networks hold informal meetings and pasture walks as social events, as an opportunity for farmers to ask questions relating to their own programs, and as a way to view how other grazers handle their pastures and rotations. These meetings are an excellent way to exchange ideas and receive support and encouragement from other grazers.

UNLIMITED RESOURCES

The resources available to you are limited only by the amount of time you dedicate to researching them. The Internet has made available huge quantities of information that may apply to your situation. As with all things related to electronic data, it is best to check the authorities of the articles you use and to use care in providing information about your farm to outside requests.

APPENDIX

BREED AND ASSOCIATION ADDRESSES

BEEF BREEDS

American Angus Association
3201 Frederick Avenue
St. Joseph, MO 64506
(816) 383-5100
www.angus.org

American Brahman Breeders Association
3003 South Loop West, Suite 140
Houston, TX 77054
(713) 713-349-0854
www.brahman.org

American British White Park Association
P.O. Box 957
Harrison, AR 72602
(877) 900-BEEF
www.whitecattle.org

American Chianina Association
1708 North Prairie View Road
P.O. Box 890
Platte City, MO 64079
(816) 431-2808
www.chicattle.org

American Dexter Cattle Association
4150 Merino Avenue
Watertown, MN 55388
(955) 446-1423
www.dextercattle.org

American Galloway Breeders Association
C/O Canadian Livestock Records Corporation
2417 Holly Lane
Ottawa, Ontario K1V 0M7
Canada
(406) 728-5719
www.americangalloway.com

American Hereford Association
P.O. Box 14059
Kansas City, MO 64101
(816) 842-3757
www.hereford.org

American Highland Cattle Association
200 Livestock Exchange Building
4701 Marion Street
Denver, CO 80216
(303) 292-9102
www.highlandcattleusa.org

American International Charolais Association
11700 NW Plaza Circle
Kansas City, MO 64153
(816) 464-5977
www.charolaisusa.com

American Red Poll Association
P.O. Box 147
Bethany, MO 64424
(660) 425-7318
www.redpollusa.com

American Shorthorn Association
8288 Hascall Street
Omaha, NE 68124
(402) 393-7200
www.shorthorn.org

American Simmental Association
One Simmental Way
Bozeman, MT 59715
(406) 587-4531
www.simmental.org

Devon Cattle Association
11035 Waverly
Olathe, KS 66061
(913) 583-1723
www.americandevon.com

North American Limousin Foundation
7383 South Alton Way, Suite 100
Englewood, CO 80112
(303) 220-1693
www.nalf.org

North American South Devon Association
19590 East Main Street, Suite 202
Parker, CO 90138
(303) 770-3130
www.southdevon.com

Pineywoods Cattle Registry and Breeders Association
2262 Highway 59
Spruce Pine, AL 35585
(256) 332-6847
www.pcrba.org

Red Angus Association of America
4201 I-35 North
Denton, TX 76207
(817) 387-3502
www.redangus.org

Santa Gertrudis Breeders International
P.O. Box 1257
Kingsville, TX 78364
(361) 592-9357
www.santagertrudis.com

DAIRY BREEDS

American Guernsey Association
7614 Slate Ridge Blvd.
Reynoldsburg, OH 43068
(614) 864-2409
www.usguernsey.com

American Jersey Cattle Association
6486 E. Main Street
Reynoldsburg, OH 43068
(614) 861-3636
www.usjersey.com

American Milking Shorthorn Society
800 Pleasant Street
Beloit, WI 53511
(608) 365-3332
www.milkingshorthorn.com

Ayrshire Breeders Association
1224 Alton Creek Road, Suite B
Columbus, OH 43228
(614) 335-0020
www.usayrshire.com

Brown Swiss Cattle Breeders Association of the USA
800 Pleasant Street
Beloit, WI 53511
(608) 365-4474
www.brownswissusa.com

Dutch Belted Cattle Association of America
c/o American Livestock Breeds Conservancy
P.O. Box 477
Pittsboro, NC 27312
(919) 542-5704
www.dutchbelted.com

Holstein Association USA
One Holstein Place
Brattleboro, VT 05302
(800) 952-5200
www.holsteinusa.com

DAIRY CATTLE BREEDING GUIDE

Animal Analysis Associates (aAa)
c/o James E. Sarbacker
602 Parkland Drive
Verona, WI 53593
(608) 848-2222
www.aaa123456.com

OTHER ORGANIZATIONS

FFA
6060 FFA Drive
Indianapolis, IN 46282
(317) 802-6060
www.ffa.org

4-H
1400 Independence Ave. S.W. Stop 2225
Washington, D.C. 20250
(202) 720-2908
www.4husa.org

INDEX